SpringerBriefs in Applied Sciences and Technology

SpringerBriefs present concise summaries of cutting-edge research and practical applications across a wide spectrum of fields. Featuring compact volumes of 50 to 125 pages, the series covers a range of content from professional to academic.

Typical publications can be:

- A timely report of state-of-the art methods
- An introduction to or a manual for the application of mathematical or computer techniques
- A bridge between new research results, as published in journal articles
- A snapshot of a hot or emerging topic
- An in-depth case study
- A presentation of core concepts that students must understand in order to make independent contributions

SpringerBriefs are characterized by fast, global electronic dissemination, standard publishing contracts, standardized manuscript preparation and formatting guidelines, and expedited production schedules.

On the one hand, **SpringerBriefs in Applied Sciences and Technology** are devoted to the publication of fundamentals and applications within the different classical engineering disciplines as well as in interdisciplinary fields that recently emerged between these areas. On the other hand, as the boundary separating fundamental research and applied technology is more and more dissolving, this series is particularly open to trans-disciplinary topics between fundamental science and engineering.

Indexed by EI-Compendex, SCOPUS and Springerlink.

Sourabhi Debnath · Tanmoy Debnath ·
Mohammad Ali Moni · Manoranjan Paul

Graphene in Wearable Sensors for Health Monitoring

Sourabhi Debnath
School of Computing, Mathematics and Engineering
Charles Sturt University
Bathurst, NSW, Australia

Tanmoy Debnath
School of Computing, Mathematics and Engineering
Charles Sturt University
Bathurst, NSW, Australia

Mohammad Ali Moni
Artificial Intelligence and Cyber Futures Institute
Charles Sturt University
Bathurst, NSW, Australia

Manoranjan Paul
School of Computing, Mathematics and Engineering
Charles Sturt University
Bathurst, NSW, Australia

ISSN 2191-530X ISSN 2191-5318 (electronic)
SpringerBriefs in Applied Sciences and Technology
ISBN 978-981-96-8849-4 ISBN 978-981-96-8850-0 (eBook)
https://doi.org/10.1007/978-981-96-8850-0

This Springer imprint is published by the registered company Springer Nature Singapore Pte Ltd.
The registered company address is: 152 Beach Road, #21-01/04 Gateway East, Singapore 189721, Singapore

Preface

This book highlights the advancement of graphene-based non-invasive sensors for health monitoring over the past decade. The story of graphene is one of both scientific ingenuity and ongoing discovery. Since its first isolation, this remarkable material has captured the attention of researchers worldwide due to its exceptional properties, ranging from its electrical conductivity and mechanical strength to its thermal and optical behaviour. What began as a novel carbon structure has evolved into a platform for innovation across disciplines, from electronics to biomedicine. With their mechanical, thermal, and electrical properties, graphene-based materials have shown effectiveness in sensing various biophysical and biochemical signals such as body temperature, heart rate, respiration rate, blood pressure, blood glucose levels and electrocardiograms, electromyograms, and electroencephalograms. This book is written to offer a clear and up-to-date account of where the field currently stands and where it may be heading.

We aim to provide a resource that is both technically robust and accessible to those with a professional interest in advanced materials. Whether the reader is a postgraduate student, an academic researcher, or an industry practitioner, we hope this book will deepen their understanding of how graphene is being developed, applied, and refined. It facilitates collaboration between key scientific foundations and highlights how graphene is integrated into real-world technologies, particularly in healthcare and sensing applications.

Chapter 1 offers a broad yet focused introduction to graphene: its structure, unique behaviours, and the breakthroughs that have defined its trajectory. From there, Chap. 2 explores how graphene is produced, comparing the top-down and bottom-up methods. While some techniques produce high-quality materials, they are often expensive and difficult to scale; others offer more accessible routes but come with compromises. This chapter discusses these trade-offs with an eye towards future development. Chapter 3 examines its biocompatibility—how it behaves inside the body, what risks and opportunities it presents, and where the current science stands.

The next chapter explores how graphene-based materials are integrated into wearable sensor technologies for health monitoring. Chapter 4 focuses on biophysical sensors and wearable devices that are transforming how we monitor our health non-invasively.

Chapter 5 takes a step back to reflect on the challenges ahead. While the potential is clear, there's a real need to bridge the gap between laboratory success and clinical reality. Issues like scalability, long-term stability, and regulatory approval still need addressing, and doing so will require close collaboration between scientists, engineers, and healthcare professionals.

The authors would like to express their sincere gratitude for the seed funding provided by the Artificial Intelligence and Cyber Futures Institute (AICF), Charles Sturt University (CSU), Australia. We also acknowledge the logistical support extended by the School of Computing, Mathematics and Engineering (SCME), CSU, which contributed significantly to the development of this work.

Bathurst, Australia
June 2025

Sourabhi Debnath
Tanmoy Debnath
Manoranjan Paul
Mohammad Ali Moni

Competing Interests The authors have no competing interests to declare that are relevant to the content of this manuscript.

Contents

Chapter 1
Introduction to Graphene

Abstract This chapter provides a focused overview of the historical development and latest scientific advances that have shaped current understanding of graphene, particularly within the context of materials science, nanotechnology, and engineering. Graphene, a monolayer of sp^2-bonded carbon atoms arranged in a two-dimensional honeycomb lattice, continues to attract significant attention due to its extraordinary intrinsic properties—high carrier mobility, quantum Hall effects, superior thermal conductivity, optical transparency, and exceptional mechanical strength. Its distinct chemical behaviour, especially in terms of asymmetric functionalisation and supramolecular interactions, further differentiates it from other carbon nanostructures. The discussion highlights ongoing research efforts addressing challenges in synthesis, scalability, defect control, and integration into functional systems. Emphasis is placed on emerging applications across electronics, advanced composites, and nanoscale devices, illustrating the convergence of fundamental research and technological development. By situating graphene within the broader landscape of two-dimensional materials, this introductory text aims to establish a framework for the in-depth analyses that follow in later chapters, while celebrating continued innovation and interdisciplinary collaboration in the field.

Keywords Graphene · Two-dimensional materials · Nanomaterials · Nanotechnology · Electrical conductivity · Thermal conductivity · Hybrid systems · Quantum hall effect · Optical property · Mechanical property · Chemical property

1.1 The History of Graphene

The history of graphene starts with the understanding that graphite is composed of numerous layers of graphene arranged on top of each other. This arrangement creates a three-dimensional structure known as graphite. In contrast, graphene, the basic building block of materials such as fullerene and carbon nanotubes, is a two-dimensional (2D) allotrope of carbon that is only one atom thick. Historically, the

S. Debnath et al., *Graphene in Wearable Sensors for Health Monitoring*,
SpringerBriefs in Applied Sciences and Technology,
https://doi.org/10.1007/978-981-96-8850-0_1

term ‘graphene’ was frequently employed to characterise the properties of various carbon allotropes. This term derives from the root word ‘graph’, indicating its connection to graphite, coupled with the suffix ‘-ene’, which denotes the presence of a carbon–carbon double bond [1].

Graphene was predicted to be unstable due to the formation of curved structures such as fullerenes, carbon nanotubes, and soot. It was also believed that it could not exist in its free state because thermodynamics had been shown to prevent the formation of two-dimensional crystals in a free state [2]. Historically, graphite was used to mark sheep and other objects, leading to its naming as ‘graphite’, derived from the Greek word ‘graphein’, meaning ‘to write’ [3]. It was first coined by the German chemist and mineralogist A.G. Werner in 1789 [4]. As the pencil industry developed, graphite has been widely used as a writing material in pencils since the eighteenth century. Due to its layered structure and weak forces between adjacent sheets, graphene has been utilised as a solid lubricant [5]. In 1947, Wallace reported the development of electronic energy bands and Brillouin zones for graphite using the ‘tight-binding’ approximation. According to the report, graphite is a semiconductor with zero activation energy, meaning there are no free electrons at absolute zero temperature because its Fermi level sits precisely at the Dirac point. However, they are generated at higher temperatures through excitation to a band adjacent to the highest one, typically filled. Assumptions about the mean free path are used to analyse electrical conductivity, which is approximately 100 times more significant, parallel to the crystal planes. A large and anisotropic diamagnetic susceptibility was anticipated for the conduction electrons, with the most significant effect observed for fields across the layers. The volume of optical absorption was also accounted for [6]. As reported in 1960, Ubbelohde and Lewis separated a single-atom layer of graphite and documented a notably increased basal-plane conductivity in graphite intercalation compounds compared to the pristine graphite.

They highlighted that graphite is composed of layers that form a network of carbon atoms arranged in hexagonal rings [7].

A specific description of graphene material has been introduced since 1986, when Boehm et al. suggested establishing a standard definition for the term: *‘The ending -ene is used for fused poly- cyclic aromatic hydrocarbons, even when the root of the name is of trivial origin, e.g., naphthalene, anthracene, tetracene, coronene, ovalene,* etc. *A single carbon layer of the graphitic structure would be the final member of an infinite series. Graphene layer should be used for such a single carbon layer’* [8, 9]. In 1994, Boehm authored the International Union of Pure and Applied Chemistry (IUPAC) report[5] where the recommendations were formalised, leading to the replacement of the term ‘graphite layers’ with ‘graphene’ in their Compendium of Chemical Technology, which states: *‘previously, descriptions such as graphite layers, carbon layers, or carbon sheets have been used for the term graphene. Because graphite designates that modification of the chemical element carbon in which planar sheets of carbon atoms, each atom bound to three neighbours in a honeycomb-like struc- ture, are stacked in a three-dimensional regular order, it is not correct to use for a single layer a term that includes the term graphite, which would imply a three-dimensional structure. The term graphene should be used only when the reactions, structural relations, or other properties of individual layers are discussed'* [10].

The first chemical vapour deposition (CVD) technique for producing monolayer graphite or graphene was used by Hess and Ban in 1966 [5]. Additionally, successful efforts have been made to grow few-layer graphite epitaxially through CVD of hydrocarbons on metal substrates [11, 12], and on top of other materials [13], as well as by thermal decomposition of SiC. Epitaxial graphene, like free-standing graphene, comprises a single layer of sp^2-bonded carbon atoms arranged in a hexagonal structure. However, there is a significant charge transfer from the substrate to the epitaxial graphene. In some instances, there is hybridisation between the d orbitals of the substrate atoms and the π orbitals of graphene, leading to a substantial alteration in the electronic structure of epitaxial graphene [5]. In 1984, Semenoff proposed that graphene could transmit electric current through massless charge carriers [14].

Between 1990 and 2004, numerous attempts were made to produce extremely thin graphite films through mechanical exfoliation; however, these efforts failed to yield samples with fewer than 50 to 100 layers within this period [5]. In 2004, Geim and Novoselov at the University of Manchester successfully isolated single-atom-thick crystallites from bulk graphite known as graphene using the 'Scotch Tape Technique' [15]. The process involved adhering multiple graphene layers to an adhesive tape, followed by repetitive folding and peeling, which reduces the material to a single layer of carbon. Acetone was used as a solvent to detach the tape, concluding with a final peel using an unutilised piece of tape onto which the graphite sample was placed. Though this process is recognised for producing high-quality graphene, scaling up this method presents challenges [5].

1.2 Structure

Graphene consists of a single layer of sp^2 hybridised carbon atoms [16] arranged in a hexagonal structure with a carbon–carbon bond and one atom thickness [1, 5]. As shown in Fig. 1.1, a stacked hexagonal configuration creates three dimensional graphite; a two-dimensional arrangement results in graphene, a rolled hexagonal formation leads to one-dimensional carbon nanotubes, and a wrapped hexagonal structure produces zero-dimensional fullerenes. The thickness of graphene is only 0.35 nm.

The bonds between carbon atoms are strong enough to withstand external forces while maintaining the integrity of the lattice structure, thereby preventing atomic reconfiguration. Graphene can also take the form of nanoribbons, which are one-dimensional strips of graphene known for their high mobility and efficient current-carrying ability [17]. In graphene nanoribbons (GNRs), the electronic structure is related to the width and boundary configuration. The energy barrier in GNRs occurs near the central point because of lateral charge movement and increases as the width of the nanoribbon decreases [18, 19].

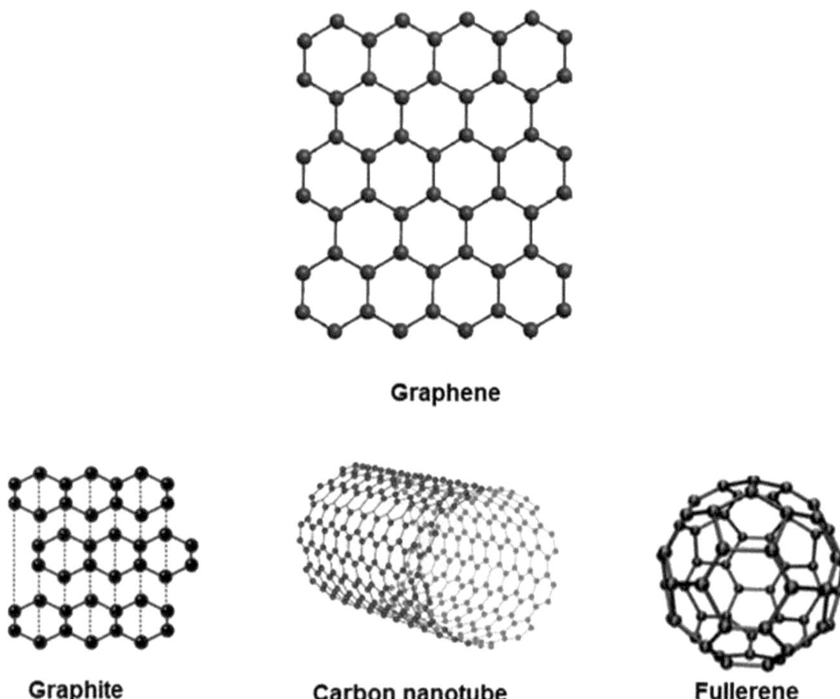

Fig. 1.1 Carbon-containing molecules (graphite, fullerene, and carbon nanotube) derived from graphene

The graphene can be classified into zigzag and armchair types based on the arrangement of carbon chains [20], as depicted in Fig. 1.2. This difference in edge configuration results in varying conductive behaviours. Early tight-binding calculations predicted that zigzag-edge GNRs typically behave like metals. In contrast, depending on its width, an armchair-edge nanoribbon can conduct electricity like a metal or a semiconductor [21, 22].

In 2006, it was reported that ab initio calculations indicated the instability of the flat band found in tight-binding calculations when spin splitting was taken into account. The current prediction suggests that zigzag GNRs have a magnetic insulating ground state characterised by ferromagnetic ordering at the edges and anti-ferromagnetic interactions across the ribbon [18, 23]. Zigzag GNRs are anticipated to demonstrate half-metallicity [24] by modifying energy levels and removing the spin state degeneracy by applying a transverse electric field in the plane [24]. Additionally, it was noted in 2006 that all armchair GNRs display semiconducting properties [18, 25].

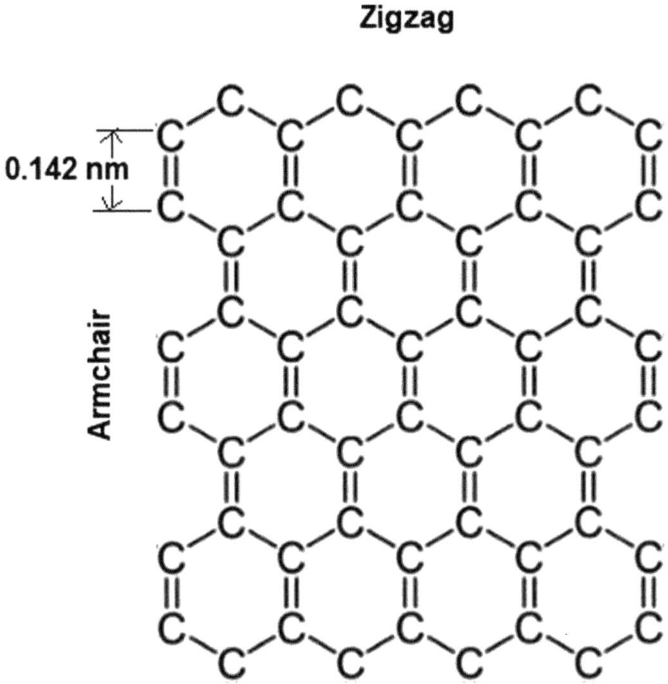

Zigzag

0.142 nm

Armchair

Fig. 1.2 Hexagonal structures of graphene with zigzag and armchair edges

1.3 Properties of Graphene

Graphene exhibits optical transparency, electrical conductivity, mechanical strength, and thermal conductivity properties. As a superlight material, its planar density stands at 0.77 mg.m^{-2}. As illustrated in Fig. 1.2, the C = C bond length is approximately 0.142 nm. The research findings indicate that a monolayer of graphene has a breaking strength of 42 N m^{-1} [26], a fracture strength of 125 GPa [27], and thermal conductivity in the range of 3000 5000 W m^{-1} K^{-1} [1, 28] and the calculated specific surface area is 2630 m^2 g^{-1} [29].

Graphene exhibits remarkable transport phenomena, such as the quantum Hall effect (QHE), and its outstanding thermal, optical, and electrical properties are attributed to its extensive π-π bonding. Moreover, among graphene that possessed surface-active functional moieties, including carboxylic, ketonic, quinonic, and C = C, the carboxylic and ketonic groups are highly reactive and easily bind with different biomolecules by forming covalent bonds. This property enhances the potential for functionalising graphene with biomolecules for different biosensing applications [30–32].

1.3.1 Electrical Property

The electronic structure of graphene can be understood using the tight-binding model. This model assumes that isolated atoms reside at each lattice point, with the amplitude of the electron wave functions diminishing as the distance from the atom increases [33, 34]. As the lattice constant is reduced, the atomic wave functions of neighbouring atoms begin to overlap. At this stage, a wave function can be proposed as a Bloch function, which must satisfy the Schr¨odinger equation [35]. Pristine graphene acts as a zero-gap semiconductor [2]. Its sp^2 hybridised carbon atoms—a combination of s, p_x, and p_y orbitals that create the sigma bond are organised in a hexagonal structure within a two-dimensional plane [19]. As shown in Fig. 1.3, each hexagonal ring features three strong in-plane σ bonds that are strongly interconnected and oriented perpendicular to the plane [35]. The final p_z electron contributes to the formation of the π bond that contributes to graphene's conductive properties [19].

Different layers of graphene are connected through weaker π interactions. The π bond is the primary source of graphene's electrical conductivity, which is oriented perpendicular to the lattice plane. Together, the π bonds combine to create π-band and π^* bands, producing valence and conduction bands. These bands are responsible for many of the electronic properties of graphene, enabled by a half-filled band that allows for the movement of free electrons.

Another characteristic of graphene is its high electron mobility. The intrinsic mobility μ of charge carriers in graphene can exceed $\sim 2 \times 10^5$ cm^2 V^{-1} s^{-1} [36]. The mobility of charge carriers is noted as 106 cm^2 V^{-1} s^{-1} [37] at $T \sim 10$ K. Graphene has the highest conductivity of any material at room temperature, with a conductivity of $5 \times 10^6 - 6.4 \times 10^6$ S m^{-1} [38]. This remarkable conductivity can be attributed to its ultrahigh mobility, which exceeds that of silicon. Due to the sp^2

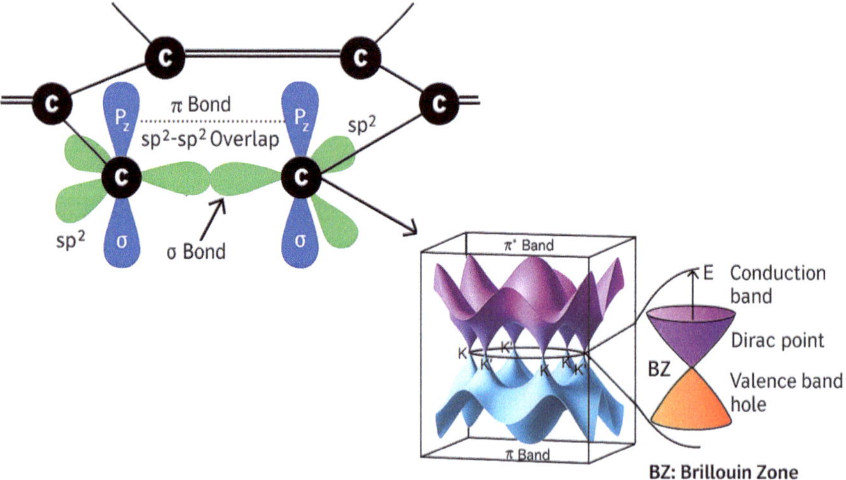

Fig. 1.3 Graphene bonds

hybridised structure, the carbon atoms in graphene donate an additional electron to the π bond. In this instance, the π electrons can move freely with minimal interference at room temperature, resulting in high conductivity [19]. Graphene also functions as a typical semimetal, displaying a slight overlap between its conduction and valence bands [35]. Electrons at the upper end of the valence band can transition to the lower end of the conduction band with reduced energy requirements, eliminating the need for thermal stimulation. Even at absolute zero temperature, a certain number of electrons exists in the conduction band along with a corresponding number of holes in the valence band [19]. Another characteristic of graphene is the unique behaviour of its charge carriers, which function as massless relativistic particles known as Dirac fermions. These Dirac fermions exhibit distinct properties, particularly in magnetic fields, including the quantum Hall effect (QHE), which is observed even at room temperature [2].

The following section provides a brief overview of the quantum Hall effect in graphene.

1.3.2 Quantum Hall Effect (QHE)

The band structure of single-layer graphene features two bands intersecting at two non-equivalent points, known as Dirac points in the Brillouin zone, where the valence and conduction bands overlap, thereby classifying graphene as a zero-band gap semiconductor [2]. In graphene, the dispersion relationship between electrons and holes is linear at the Dirac points, i.e. behaving similarly to relativistic particles, indicating that the effective mass of both electrons and holes is zero [34].

The integer quantum Hall effect (IQHE), first identified in 1980 [39], is a phenomenon observed in two-dimensional electron systems, characterised by quantised Hall resistance in the perpendicular direction to the current at low temperatures and under strong magnetic fields. The quantisation is expressed as $R_H = \frac{h}{}$ and depends on the charge of the electron (e) and Planck's constant (h), where i is an integer value with i = (1, 2, 3,...). This effect can be understood through Landau quantisation, which reveals that electrons in a strong magnetic field within a two-dimensional framework exhibit a discrete energy spectrum characterised by highly degenerate energy levels, known as Landau levels (LLs). The fractional quantum Hall effect (FQHE), discovered in 1982, expanded the understanding of these phenomena by demonstrating that the filling factor can be fractional rather than just an integer [40]. This discovery earned Daniel C. Tsui, Horst L. St¨ormer, and Robert B. Laughlin the Nobel Prize in Physics in 1998.

The relativistic quantum Hall effect in graphene was first identified in 2005 [34]. Graphene's hexagonal lattice comprises two interpenetrating sublattices and preserves inversion symmetry. This topological characteristic gives rise to its unique conical energy dispersion. When analysing the relativistic quantum Hall effect in graphene, it is essential to consider both charge carriers: electrons and holes. The challenge arises from the confinement potential, which limits the use of standard methods for the classical integer Hall effect. If the confinement potential approaches infinity, electrons can be confined; however, holes, which carry opposite charges, would remain unconfined or conversely, if the potential becomes negatively infinite [33]. The solution employs a mass confinement potential defined in terms of Pauli matrices [39].

Generally, the quantum electronic characteristics that govern the two-dimensional electron systems are described using the Schr¨odinger equation, implying that quasi-particles behave in a non-relativistic manner with a well-defined effective mass [39]. However, graphene electrons act as massless relativistic particles and follow the Dirac equation of motion within two dimensions [41–43]. Of particular interest are the energy levels near the Dirac points, denoted as K and K' (Fig. 1.3) exhibiting linear dispersion concerning the wave vector, and hence, the electrons always travel at a constant speed with the band velocity vF = 106 s^{-1} and indicating that the effective mass of electrons and holes is zero [2, 34].

Similar to LLs in conventional two-dimensional electron systems characterised by quadratic band dispersion, the LLs in graphene are also highly degenerate, with the number of states per LL being determined by the total count of flux quanta penetrating the two-dimensional surface [44]. Standard two-dimensional electron systems typically exhibit twofold spin degeneracy, independent of the Zeeman effect. In contrast, the LLs in graphene manifest a fourfold degeneracy, principally driven by the dominant Coulomb repulsion that respects twofold spin degeneracy and twofold sublattice symmetry, transcending the conventional spin degeneracy[45][45]. This additional degeneracy can be attributed to the twofold valley degeneracy arising from the inequivalent K and K' points, where the Fermi level intersects the electronic bands [47]. As a result, there is a possibility that electrons can create sa quantum–mechanical superposition of the orbital wave functions corresponding to each of the four spin-valley combinations

$$|\uparrow, K\rangle, \rightarrow |\uparrow, K'\rangle, \rightarrow |\downarrow, K\rangle, \rightarrow |\downarrow, K'\rangle$$

where ↑,↓ represents the spin orientation[44].

In graphene, the relativistic quantum Hall effect demonstrates that rather than experiencing plateaus with quantised resistance at integer values of n, plateaus are observed at n = ± 2(2 m + 1), where m is an integer and represents the LL index m = 0,1,2… [48] as confirmed by experimental studies[49, 50]. This results in plateaus at values such as ± 2, ± 6, ± 10, and so on. The presence of positive and negative values in this sequence indicates that two types of charge carriers—electrons and holes—contribute to the QHE. The characteristics of these carriers can be modified through the electric field effect [33]. This sequence of n highlights the relativistic characteristics of electrons in graphene. The characteristics of these carriers can be modified through the electric field effect. Many of the fascinating electronic properties of graphene derive from this dispersion relationship, which resembles that of relativistic, massless fermions.

1.3.3 Thermal Conductivity

The thermal transport in graphene is predominantly phonon-dominated [5] and its thermal characteristics draw on graphite, highlighting this material's highly anisotropic nature [51]. The 2D structure of graphene includes three acoustic phonon modes: the in-plane longitudinal (LA), transverse (TA), and out-of-plane flexural (ZA) branches with a quadratic dispersion relation [5, 52]. This high in-plane thermal conductivity results from covalent sp^2 bonding between carbon atoms, which has a bonding energy estimated at approximately 5.9 eV [51]. In contrast, the weak van der Waals forces within a graphite crystal, characterised by a spacing of about 3.35 Å, limit heat flow in the out-of-plane direction [51]. When compared to other carbon allotropes, such as multi-walled carbon nanotubes (MWCNTs) and single-walled carbon nanotubes (SWCNTs), which possess thermal conductivities approximately 3000 Wm^{-1} K^{-1} and 3500 Wm^{-1} K^{-1} respectively, pure single-layer graphene demonstrates a significantly greater thermal conductivity of approximately 5000 Wm^{-1} K^{-1} at 27 °C, attributed to its in-plane carbon bonds, making it an outstanding conductor of heat [5].

Historically, it was widely accepted that in-plane acoustic phonons predominantly facilitated thermal transport in graphene. However, in 2007, [53] assessed the ballistic thermal conductance of graphene by examining the phonon and electron dispersion relations. This research demonstrated that, below approximately 20 K, the ballistic conductance is primarily governed by the out-of-plane flexural (ZA) mode. In contrast, contributions from the LA and TA modes become relevant at temperatures exceeding 20 K [53]. Moreover, Seol et al. conducted comprehensive experiments on the thermal transport properties of single-layer graphene (SLG) on amorphous SiO_2. Subsequently, they reassessed the thermal conductivity of suspended graphene. Their findings indicated that at 300 K, the ZA branch could account for up to 77% of the thermal conductivity, increasing to 86% at 100 K, attributable to its high specific heat capacity and extended mean scattering time [54]. Further investigations by Zhang et al. demonstrated that in GNR systems, the ZA branch exhibits a considerably higher thermal conductivity than the LA and TA branches [55–57]. Their research also highlighted that energy transfer among ZA phonons transpires significantly faster than the energy transfer between ZA, LA, and TA phonons. It has been established that flexural mode phonons can effectively dissipate heat more rapidly than their longitudinal or transverse counterparts, regardless of whether they are subjected to a mobile or stationary localised heat source.

Notably, when graphene is positioned on a substrate or confined within GNRs, its in-plane thermal conductivity experiences a substantial reduction compared to samples that are freely suspended [51]. This decrease can be anticipated considering the sensitivity of phonon propagation within a monolayer graphene sheet to disruptions caused by surfaces or edges. Empirical measurements indicate that at room temperature, the thermal conductivity of graphene supported on SiO_2 is approximately 600 W m^{-1} K^{-1}, while SiO_2-encased graphene exhibits a decreased thermal conductivity of approximately 160 W m^{-1} K^{-1}. Furthermore, supported GNRs, especially those with widths approximating 20 nm, display a thermal conductivity estimated at roughly 80 W m^{-1} K^{-1} [51].

When confined in GNRs with widths smaller than the intrinsic phonon mean free path, phonon scattering at boundaries and edge roughness leads to even lower thermal conductivity than that observed in suspended or SiO_2-supported graphene [58, 59]. Moreover, intrinsic ripples in graphene demonstrate inharmonic interactions between bending and stretching modes, which help maintain long-range order in 2D crystals [52]. As these ripples develop into more prominent wrinkles, the mean bond length in graphene increases, leading to a significant rise in the phonon density of states. The chirality-dependent bending stiffness and disparities in wave velocity contribute acutely to transverse wave propagation in GNRs, which are sensitive to the excitation frequency of the vibrational source and the chirality direction [60]. The minimum permissible wavelength is quantified as 3.69 Å in the zigzag direction and 5.68 Å in the armchair direction [60]. Consequently, adjusting the level and orientation of wrinkles along various chiral pathways modulates the thermal conductivity in graphene. Defects, edge scattering, and isotopic doping have a significant influence on thermal conductivity [5]. These elements negatively impact conductivity through the introduction of phonon scattering at defect sites, as well as the localisation of phonon modes due to doping. Structural defects, including non-hexagonal rings and vacancies, considerably affect the anisotropic nature of thermal conductivity, as defect regions serve as effective scattering centres, particularly when present in minimal quantities [52].

1.3.4 Optical Property

Graphene is recognised for its transparency, making it suitable for various photonic devices that require thin yet transparent films. A single graphene layer absorbs approximately 2.3% of white light with an optical transmittance of 97% [61]. The fine-structure constant can accurately characterise the transmittance of graphene. As additional layers are added, light absorption increases proportionally; each layer demonstrates absorption of $A = 1—T = \pi\alpha = 2.3\%$, where $\alpha = 1/37$ denotes the fine-structure constant [2]. Therefore, its stacking order and orientation influence the optical properties of graphene. Graphene can be visualised through optical image contrast on a Si/SiO_2 substrate, which relies on interference. The refraction and interference of light through graphene of varying thicknesses produce distinct colours and contrasts [62]. Both theoretical models and experimental results showed that the optical properties can be adjusted by altering the thickness. Visualisation of graphene on a Si/SiO_2 substrate can be accomplished through optical image contrast generated by interference effects, which significantly enhances with increasing layer thickness [63]. The graphene's absorption spectrum remains stable across the 300–2500 nm range, featuring a pronounced peak around 250 nm in the ultraviolet region, associated with electronic transitions from the unoccupied π-states [64]. Given its high transparency and low resistivity, graphene is an exemplary candidate for electrodes in liquid crystal displays [65].

Due to its exceptional conductivity, graphene shows excellent promise as a transparent and conductive membrane [66, 67]. Moreover, with specific modifications such as doping, it can function as an acceptor in optoelectronic devices [68, 69] and serve as an electrode in supercapacitors [70]. Additionally, graphene exhibits saturated light absorption when light intensity surpasses a certain critical threshold, especially in the near-infrared range, owing to its broad absorption capabilities and zero band gap resulting from the linear dispersion of Dirac electrons in graphene, making it an attractive candidate for ultrafast mode-locked lasers [19, 71, 72]. Nonequilibrium carriers and chemical or physical treatments of graphene can also lead to luminescence. These properties make it an ideal photonic and optoelectronic material [73–78].

The transparency of graphene layers was systematically compared to that of indium tin oxide (ITO) glasses in [79]. The researchers synthesised graphene layers through the pyrolysis of varying quantities of camphor, and their optical and electrical evaluations revealed a sheet resistance of 860 $\omega.cm^{-2}$ and a transmittance of 91% at a wavelength of 550 nm for the graphene film. Notably, the transmittance remained above 80% in the range from 250 to 1750 nm, surpassing the performance of ITO glass, which has a transmission range of 250 nm to 800 nm. ITO has long been employed as a conductor for electrons and holes in numerous applications, attributed to its high transparency (approximately 90%) to visible light, low sheet resistance (10–30 $\omega.sq^{-1}$), and favourable work function (4.5–5.2 eV). However, ITO films have limitations, such as being unable to accommodate the deposition of semiconducting materials at high temperatures unless prepared on quartz glass, which drives up costs. ITO films are also unstable under specific pH conditions, which limits their usability [64, 80].

1.3.5 Mechanical Property

Graphene sheets exhibit remarkable flexibility and can be stretched while withstanding pressure differences in several atmospheres. The exceptional strength in graphene is primarily due to the covalent sp^2 hybridisation among carbon atoms. Graphene weighs approximately 0.77 mg.m^{-2} [64].

It has Young's Modulus of 1500 GPa [81] and tensile strength of 130 GPa [82], making it the strongest known material and could support a load of 13 tonnes across an area of 1 mm^2 [83].

The mechanical characteristics of crystalline solids are primarily determined by the properties of their ideal lattice structure and structural defects, e.g. dislocations and grain boundaries. The interactions between atoms and the lattice's geometry influence the material's elastic properties in a defect-free crystal lattice [84]. The presence of defects, such as dislocations and grain boundaries, can significantly reduce the strength of real materials under mechanical stress compared to the theoretical strength of defect-free materials [85].

A fundamental study conducted by Griffith indicated that the true breaking strength of a brittle material is determined by the size of defects and imperfections rather than the inherent strength of its atomic bonds [86]. When a material's size approaches or falls below 102 nm, the defect dynamics deviate considerably from those observed in macroscopic systems [87]. The calculated elasticity values align with experimental results and only apply without defects. Nevertheless, graphene materials possess structural imperfections.

While structural defects are generally present in conventional macroscale solids and critically impact their mechanical properties, nanoscale solids may lack defects in their initial, undeformed state due to nanoscale and free-surface effects [88]. This absence of defects facilitates exceptional strength close to their ideal highest value, particularly in more miniature graphene sheets or membranes synthesised in a pristine, defect-free condition [85].

Lee et al. conducted an extensive experimental analysis of the elastic properties and strength of pristine graphene [26]. In their investigation, a graphene membrane was mechanically deposited onto a substrate featuring arrays of circular wells, which was subsequently subjected to loading using an atomic force microscope tip. Their findings illustrated that graphene exhibits nonlinear elastic behaviour and brittle fracture. The nonlinear elastic response to tensile stress was formulated by the equation $\Sigma = E\epsilon + D\epsilon^2$, where Σ denotes the applied stress (symmetric second Piola–Kirchhoff stress), ϵ signifies the elastic strain (uniaxial Lagrangian strain), E represents Young's modulus, and D indicates the third-order elastic stiffness. The results revealed that graphene possesses Young's modulus of E = 1.0 TPa and a third-order elastic stiffness of D = -2.0 TPa. Additionally, theoretical calculations based on ab initio density functional theory (DFT) also predicted the exceptional strength of graphene [87, 89]. In this context, strain is mathematically defined as $\epsilon = L/L_0 - 1$. DFT calculations suggest that graphene behaves isotropically under minimal strains (ϵ ¡ 0.1), yielding an estimated Young's modulus of approximately 1050 GPa, with Poisson's ratio of 0.186 [87]. The simulations were conducted using a four-atom cell. Notably, as ϵ exceeds 0.1, the symmetry of the lattice was disrupted, resulting in an anisotropic stress response contingent upon the direction of strain. The expected maximum tensile strength along the zigzag (armchair) direction in graphene is 121 (110) GPa at ϵ values of 0.266 (0.194). The calculated values align closely with the experimental findings. However, it is crucial to emphasise that the elasticity values reported for graphene are valid only in the absence of defects. Generally, graphene materials contain some structural imperfections [90–92]. It is well-established that the presence of substantial flaws, including monovacancies, Stone–Wales dislocations, slits, and holes, impact the mechanical properties of graphene [93]. For instance, the introduction of considerable slits or holes can reduce the fracture strength of graphene sheets to values as low as 30 to 40 GPa [83, 93, 94].

Table 1.1 summarises the different properties of graphene, and in Table 1.2, the thermal conductivity and Young's modulus of different semiconductors are compared with that of graphene.

Table 1.1 Properties of graphene

Attributes	Details
Young's modulus	1500 GPa [81]
Breaking strength	42 N m^{-1} [26]
Tensile strength	130 GPa [82]
Fracture strength	125 GPa [27]
Specific surface area	~ 2630 m^2 g^{-1}[29]
Thermal conductivity	3000–5000 W m^{-1} K^{-1} [28]
Thermal stability	450–650 °C [95]
Electron mobility	~ 2 × 10^5 cm^2 V^{-1} s^{-1} [36]
Electrical conductivity	5 × 10^6–6.4 × 10^6 S m^{-1}[38]
Resistivity	50 $\mu\Omega$ cm (in-plane) [95]
Optical transmittance	~ 97%

Table 1.2 Comparison of properties of graphene and different semiconductors

Material	Thermal conductivity (Wm^{-1} K^{-1})	Young's modulus (GPa)
Graphene	3000–5000 [28]	1500 [81]
Si	139.4 ± 1.4 [97]	131 [81]
SiC	410 [81]	450 [81]
Ge	60 [81]	103 [81]
ITO	10.2 [98]	~ 190 [99]
CNT	3500 [81]	1000 [81]
GaAs	55 [100]	85.5 [101]
Diamond	1000–2200 [81]	1050–1200 [81]

1.3.6 Chemical Property

The single-atom-thick 2D structure, large specific surface area, and remarkable properties lead graphene sheets to exhibit chemical behaviours distinct from other nanocarbons, particularly regarding their asymmetric modification and unique, appealing supramolecular chemistry [102, 103]. The inherent chemical inactivity of graphene has enabled it to protect metals and metal alloys from oxidation [5]. Pioneering research from 2011 demonstrated graphene's resistance to oxidation by applying a graphene coating to copper and copper-nickel substrates using chemical vapour deposition (CVD) methods [104]. However, pristine graphene poses challenges since it is insoluble, difficult to manipulate, and decomposes before it can melt [105]. It typically exhibits poor catalytic properties [106, 107] and weak interactions with small molecules or polymers [108, 109]. It hinders its use in catalysis, sensors, and composites and makes conventional processing methods ineffective for shaping it.

It was found that polycyclic aromatic hydrocarbons (PAHs) can be used to disperse graphene [110] in the solvent similar to the dispersion of SWCNTs [111]. Likewise, in SWCNTs [112], research has shown that larger PAHs have a greater binding energy for graphene than smaller ones [113]. However, the solubilisation of larger PAHs in the solvent could be challenging [114]. Therefore, modifications to graphene, such as the development of graphene derivatives, functionalisation, doping, photochemistry, catalytic chemistry, and supramolecular chemistry, are often necessary to meet the demands of practical applications [103, 109].

1.3.6.1 Functionalisation

Chemical functionalisation can effectively modify graphene's optical, chemical, and mechanical properties [105, 115–117]. The chemical reactivity of the graphene basal plane is inherently limited due to the extensive π-conjugation, minimal structural curvature, and absence of dangling bonds [105, 118, 119]. It is intrinsically feasible for covalent addition to transform the carbon atoms from sp^2 to sp^3 hybridisation [103]. The planar aromatic carbons change to a tetrahedral geometry with longer bonds during this process. This functionalisation induces a change in the shape of the carbon atoms, resulting in a significant increase in the amount of energy required to complete the process. The necessity for high-energy reactants, such as hydrogen atoms, fluorine atoms, strong acids, and radicals, becomes apparent in overcoming these barriers to facilitate the chemical transformation [103, 120]. Graphene sheets can be uniformly dispersed in both aqueous and organic media through the selective attachment of functionalities to their surfaces [120–123].

1.3.6.2 Graphene Oxide (GrO)

Pristine graphene is highly hydrophobic [109]. Graphene oxide (GrO), a graphene derivative, exhibits solubility in both polar and non-polar solvents [124]. GrO sheets function as amphiphilic macromolecules, possessing hydrophilic edges and a hydrophobic basal plane [125, 126]. An investigation into the origins of chemical and kinetic stability in GrO revealed that thermal reduction becomes inefficient at moderate temperatures ($\sim$706 °C), resulting in a metastable state following synthesis [127].

This research employed first-principles calculations and statistical methods to investigate the low-temperature decomposition processes and ageing effects on the structure and stability of GrO. Their findings presented that the stability of GrO is affected by the tendency of oxygen functionalities to cluster, generating highly oxidised regions intermixed with pristine graphene areas.

Table 1.3 Properties of graphene, GrO, and rGrO

Property	Graphene	Graphene oxide (Gr0)	Reduced graphene oxide (rGrO)
Electron mobility	~ 2×10^5 $cm^2\ V^{-1}\ s^{-1}$ [36]	0.05–200 $cm^2\ V^{-1}\ s^{-1}$ [129]	372 $cm^2\ V^{-1}\ s^{-1}$[130]
Thermal conductivity	3000–5000 W $m^{-1}\ K^{-1}$ [28]	72 W $m^{-1}\ K^{-1}$ [131]	1390 ± 65 W $m^{-1}\ K^{-1}$ [132]
Young's modulus	1500 GPa [81]	207.6 ± 23.4 GPa [133]	1.62 GPa [134]
Specific surface area	~ 2630 $m^2\ g^{-1}$ [29]	736.6 $m^2\ g^{-1}$ [135]	758 $m^2\ g^{-1}$ [136]
Optical transmittance	~ 97% [96]	24.8%[137]	36% [138]
Solubility in water	Not dispersible [105]	6.6 μg mL^{-1}[139]	4.74 μg mL^{-1} [139]

1.3.6.3 Reduced Graphene Oxide (rGrO)

Chemically functionalised reduced graphene oxide (rGrO) similarly exhibits molecular characteristics, enabling the assembly of macroscopic materials with tailored compositions and microstructures through hydrogen bonding, hydrophobic interactions, π π stacking, and/or electrostatic interactions among graphene sheets [103, 128]. This characteristic enables the adjustment of graphene's conductivity by selecting appropriate molecules for surface adsorption. Table 1.3 presents comparisons between graphene and its derivatives.

1.3.6.4 Doping

Hetero-atom doping can modify graphene's electronic structure and intrinsic properties [140, 141]. Doping opens a band gap near the Fermi level and transforms graphene from a 'metallic' to a 'semiconducting' material [142, 143]. By covalently attaching electron-donating or withdrawing groups, researchers can create p-type or n-type materials with distinct electrical, magnetic, and optical characteristics, presenting them as highly promising for super capacitors, catalysis, and energy storage systems applications [144–147].

It was found from the theoretical study that the edges of graphene sheets influence their chemical properties as well as reactivity [148, 149]. The edge sites of a graphene sheet, which have dangling bonds, are more reactive than the basal plane [5, 150–152]. These dangling bonds can covalently bond with different chemical moieties. Such chemical moieties may enhance the solubility and processability of graphene or introduce reactive groups for further modification. Covalent functionalisation of the basal plane disrupts the π π conjugation system, whereas non-covalent functionalisation preserves graphene's intrinsic atomic and electronic architecture [103]. Additionally, the number of layers and the stacking order influence graphene's electronic properties and reactivity, as these properties vary according to the atomic arrangement of adjacent layers. The absence of van der Waals forces with the top carbon atoms yields distinct environments for surface electrons compared to those in the second layer, thereby causing variations in electronic interactions [5]. Thus, the number of layers and the stacking type also affect the reactivity of graphene [153, 154]. It has also been reported that a single graphene sheet is approximately ten times more reactive than a bilayer or multilayer graphene [155].

1.4 Summary

Graphene, a one-atom-thick layer of carbon, is gaining increasing attention in the science and engineering sectors due to its distinct physicochemical properties, mechanical strength, and considerably high electrical and thermal conductivities compared to other semiconductors. This chapter provides a brief overview of graphene's history, structure, and various properties and presents a comparison of its thermal conductivity and Young's modulus with those of different semiconductors. Furthermore, a comparison of other properties of graphene and its derivatives is also presented.

There are various methods for producing graphene and its derivatives, including epitaxial growth, exfoliation, and CVD, which will be presented in the next chapter.

References

1. N.A.A. Ghany, S.A. Elsherif, H.T. Handal, Revolution of graphene for different applications: State-of-the-art. Surfaces and Interfaces **9**, 93–106 (2017)
2. S. Alwarappan, A. Kumar, Graphene-based materials: science and technology. Taylor & Francis, Abingdon-on-Thames, UK (2013). https://books.google.com.au/books?id=xUfGAAAAQBAJ
3. Encyclopædia Britannica: Graphite (2023). https://www.britannica.com/science/graphite. Accessed XX 10 2023
4. A. G. Werner, Kurze Klassifikation und Beschreibung der Verschiedenen Gebirgsarten. Leipzig, Germany (1789)
5. M. Sharon, M. Sharon, H. Shinohara, A. Tiwari, Graphene: an introduction to the fundamentals and industrial applications. *Advanced Material Series*, Wiley, Hoboken, NJ, USA (2015). https://books.google.com.au/books?id=70ZECgAAQBAJ
6. P.R. Wallace, The band theory of graphite. Phys. Rev. **71**(9), 622 (1947)
7. A.R. Ubbelohde, F.F. Lewis, The conductivity of graphite and its intercalation compounds, in *Proceedings of the Royal Society of London. Series A, Mathematical and Physical Sciences*, vol. 254 no. 1277, pp. 185–198 (1960). https://doi.org/10.1098/rspa.1960.0012
8. H.P. Boehm, R. Setton, E. Stumpp, Nomenclature and terminology of graphite intercalation compounds (iupac recommendations 1994). Pure Appl. Chem. **66**(9), 1893–1901 (1994)
9. C. Berger, Z. Song, T. Li, X. Li, A.Y. Ogbazghi, R. Feng, Z. Dai, A.N. Marchenkov, E.H. Conrad, P.N. First et al., Ultrathin epitaxial graphite: 2d electron gas properties and a route toward graphene-based nanoelectronics. J. Phys. Chem. B **108**(52), 19912–19916 (2004)

10. E. Fitzer, K.-H. Kochling, H. Boehm, H. Marsh, Recommended terminology for the description of carbon as a solid (iupac recommendations 1995). Pure Appl. Chem. **67**(3), 473–506 (1995)
11. T. Land, T. Michely, R. Behm, J. Hemminger, G. Comsa, Stm investigation of single layer graphite structures produced on pt (111) by hydrocarbon decomposition. Surf. Sci. **264**(3), 261–270 (1992)
12. A. Nagashima, K. Nuka, H. Itoh, T. Ichinokawa, C. Oshima, S. Otani, Electronic states of monolayer graphite formed on tic (111) surface. Surf. Sci. **291**(1–2), 93–98 (1993)
13. C. Oshima, A. Nagashima, Ultra-thin epitaxial films of graphite and hexagonal boron nitride on solid surfaces. J. Phys.: Condens. Matter **9**(1), 1 (1997)
14. G.W. Semenoff, Condensed-matter simulation of a three-dimensional anomaly. Phys. Rev. Lett. **53**(26), 2449 (1984)
15. V.B. Mbayachi, E. Ndayiragije, T. Sammani, S. Taj, E.R. Mbuta et al., Graphene synthesis, characterization and its applications: A review. Results in Chemistry **3**, 100163 (2021)
16. A. Bianco, H.-M. Cheng, T. Enoki, Y. Gogotsi, R.H. Hurt, N. Koratkar, T. Kyotani, M. Monthioux, C.R. Park, J.M. Tascon et al., All in the graphene family–a recommended nomenclature for two-dimensional carbon materials. Elsevier (2013)
17. C. Tian, W. Miao, L. Zhao, J. Wang, Graphene nanoribbons: current status and challenges as quasi-one-dimensional nanomaterials. Reviews in Physics **10**, 100082 (2023)
18. Y.-W. Son, M.L. Cohen, S.G. Louie, Energy gaps in graphene nanoribbons. Phys. Rev. Lett. **97**(21), 216803 (2006)
19. H. Zhu, Graphene: fabrication, characterizations, properties and applications. Academic Press, Cambridge, MA (2017). https://books.google.com.au/books?id=E-5GDgAAQBAJ
20. J. Kim, N. Lee, Y.H. Min, S. Noh, N.-K. Kim, S. Jung, M. Joo, Y. Yamada,Distinguishing zigzag and armchair edges on graphene nanoribbons by x-ray photoelectron and raman spectroscopies. ACS Omega **3**(12), 17789–17796 (2018). PMID: 31458375. https://doi.org/10.1021/acsomega.8b0274410.1021/acsomega.8b02744
21. K. Nakada, M. Fujita, G. Dresselhaus, M.S. Dresselhaus, Edge state in graphene ribbons: nanometer size effect and edge shape dependence. Phys. Rev. B **54**(24), 17954 (1996)
22. K. Wakabayashi, M. Fujita, H. Ajiki, M. Sigrist, Electronic and magnetic properties of nanographite ribbons. Phys. Rev. B **59**(12), 8271 (1999)
23. J.J. Palacios, J. Fern´andez-Rossier, L. Brey, Vacancy-induced magnetism in graphene and graphene ribbons. Phys. Rev. B—Condensed Matter Mater. Phys. **77**(19), 195428 (2008)
24. Y.-W. Son, M.L. Cohen, S.G. Louie, Half-metallic graphene nanoribbons. nature **444**(7117), 347–349 (2006)
25. V.B.O.H.G.E. Scuseria, Electronic structure and stability of semiconducting graphene nanoribbons. Nano Lett **6**, 2748–2006 (2006)
26. C. Lee, X. Wei, J.W. Kysar, J. Hone, Measurement of the elastic properties and intrinsic strength of monolayer graphene. Science **321**(5887), 385–388 (2008)
27. J. Wang, Z. Li, G. Fan, H. Pan, Z. Chen, D. Zhang, Reinforcement with graphene nanosheets in aluminum matrix composites. Scripta Mater. **66**(8), 594–597 (2012)
28. D. Li, T. Li, Z. Mao, Y. Zhang, B. Wang, Heat transfer mechanism in graphene reinforced peek nanocomposites. RSC Adv. **13**(39), 27599–27607 (2023)
29. B. Shubha, B. Praveen, V. Bhat, Graphene-calculation of specific surface area. Int. J. Appl. Eng. Manag. Lett. **7**, 91–97 (2023)
30. M. Ioni¸t˘a, G.M. Vl˘asceanu, A.A. Watzlawek, S.I. Voicu, J.S. Burns, H. Iovu, Graphene and functionalized graphene: extraordinary prospects for nano bio composite materials. Compos. Part B: Eng. **121**, 34–57 (2017)
31. S. Shahriari, M. Sastry, S. Panjikar, R. SinghRaman, Graphene and graphene oxide as a support for biomolecules in the development of biosensors. Nanotechnol., Sci. Appl., 197–220 (2021)
32. D. Chen, H. Feng, J. Li, Graphene oxide: preparation, functionalization, and electrochemical applications. Chem. Rev. **112**(11), 6027–6053 (2012)
33. M.O. Goerbig,Quantum hall effects (2009). arXiv:0909.1998
34. S. Taylor, Quantumhall effect (2015)

35. L.E.F.F. Torres, S. Roche, J.C. Charlier, Introduction to Graphene-Based Nanomaterials: From Electronic Structure to Quantum Transport. Cambridge University Press, Cambridge, UK (2014). https://books.google.com.au/books?id=y05kAgAAQBAJ
36. S. Mao, H. Pu, J. Chen, Graphene oxide and its reduction: modeling and experimental progress. RSC Adv. **2**(7), 2643–2662 (2012)
37. E.V. Castro, H. Ochoa, M. Katsnelson, R. Gorbachev, D. Elias, K. Novoselov, A. Geim, F. Guinea, Limits on charge carrier mobility in suspended graphene due to flexural phonons. Phys. Rev. Lett. **105**(26), 266601 (2010)
38. J. Krupka, W. Strupinski, Measurements of the sheet resistance and conductivity of thin epitaxial graphene and sic films. Appl. Phys. Lett. **96**(8) (2010)
39. M.O. Goerbig, The quantum hall effect in graphene–a theoretical perspective. C R Phys. **12**(4), 369–378 (2011)
40. D.C. Tsui, H.L. St¨ormer, A.C. Gossard, Two-dimensional magnetotransport in the extreme quantum limit. Phys. Rev. Lett. **48**(22), 1559–1562 (1982) https://doi.org/10.1103/PhysRevLett.48.1559
41. J. Falkenbach,Solving the dirac equation in a two-dimensional spacetime background with a kink. Ph.D thesis, Massachusetts Institute of Technology (2005)
42. M.I. Katsnelson, Graphene: carbon in two dimensions. Mater. Today **10**(1–2), 20–27 (2007)
43. Z. Jiang, Y. Zhang, Y.-W. Tan, H. Stormer, P. Kim, Quantum hall effect in graphene. Solid State Commun. **143**(1–2), 14–19 (2007)
44. M.O. Goerbig, From the integer to the fractional quantum hall effect in graphene (2022). arXiv:2207.03322
45. L.-A. Wu, M. Murphy, M. Guidry, So (8) fermion dynamical symmetry and strongly correlated quantum hall states in monolayer graphene. Phys. Rev. B **95**(11), 115117 (2017)
46. Y. Barlas, K. Yang, A. MacDonald, Quantum hall effects in graphene-based two-dimensional electron systems. Nanotechnology **23**(5), 052001 (2012)
47. Y. Zhang, Electronic transport in graphene. PhD thesis, Columbia University (2006)
48. C.R. Dean, A.F. Young, P. Cadden-Zimansky, L. Wang, H. Ren, K. Watanabe, T. Taniguchi, P. Kim, J. Hone, K. Shepard, Multicomponent fractional quantum hall effect in graphene. Nat. Phys. **7**(9), 693–696 (2011)
49. K.S. Novoselov, A.K. Geim, S.V. Morozov, D. Jiang, M.I. Katsnelson, I.V. Grigorieva, S.V. Dubonos, A.A. Firsov, Two-dimensional gas of massless dirac fermions in graphene. Nature **438**(7065), 197–200 (2005)
50. Y. Zhang, Y.W. Tan, H.L. Stormer, P. Kim, Experimental observation of the quantum hall effect and berry's phase in graphene. Nature **438**(7065), 201–204 (2005)
51. E. Pop, V. Varshney, A.K. Roy, Thermal properties of graphene: Fundamentals and applications. MRS Bull. **37**(12), 1273–1281 (2012)
52. J. Zhang, F. Xu, Y. Hong, Q. Xiong, J. Pan, A comprehensive review on the molecular dynamics simulation of the novel thermal properties of graphene. RSC Adv. **5**(109), 89415–89426 (2015)
53. K. Saito, J. Nakamura, A. Natori, Ballistic thermal conductance of a graphene sheet. Phys. Rev. B—Condensed Matter Mater. Phys. **76**(11), 115409 (2007)
54. J.H. Seol, I. Jo, A.L. Moore, L. Lindsay, Z.H. Aitken, M.T. Pettes, X. Li, Z. Yao, R. Huang, D. Broido et al., Two-dimensional phonon transport in supported graphene. Science **328**(5975), 213–216 (2010)
55. J. Zhang, X. Wang, H. Xie, Phonon energy inversion in graphene during transient thermal transport. Phys. Lett. A **377**(9), 721–726 (2013)
56. J. Zhang, X. Wang, H. Xie, Co-existing heat currents in opposite directions in graphene nanoribbons. Phys. Lett. A **377**(41), 2970–2978 (2013)
57. J. Zhang, X. Wang, Thermal transport in bent graphene nanoribbons. Nanoscale **5**(2), 734–743 (2013)
58. A. Bagri, S.-P. Kim, R.S. Ruoff, V.B. Shenoy, Thermal transport across twin grain boundaries in polycrystalline graphene from nonequilibrium molecular dynamics simulations. Nano Lett. **11**(9), 3917–3921 (2011)

59. Z. Aksamija, I. Knezevic, Lattice thermal conductivity of graphene nanoribbons: Anisotropy and edge roughness scattering. Appl. Phys. Lett. **98**(14) (2011)
60. X. Liu, F. Wang, H. Wu, Anisotropic propagation and upper frequency limitation of terahertz waves in graphene. Appl. Phys. Lett. **103**(7) (2013)
61. D.E. Sheehy, J. Schmalian, Optical transparency of graphene as determined by the fine-structure constant. Phys. Rev. B—Condensed Matter Mater. Phys. **80**(19), 193411 (2009)
62. K. Kim, Y. Zhao, H. Jang, S.Y. Lee, J.M. Kim, J.H. Ahn, Large-scale pattern growth of graphene films for stretchable transparent electrodes. Nature **320**, 2758–2763 (2008)
63. S. Roddaro, P. Pingue, V. Piazza, V. Pellegrini, F. Beltram, The optical visibility of graphene: Interference colors of ultrathin graphite on sio_2. Nano Lett. **7**, 2707–2710 (2007). https://doi.org/10.1021/nl0711581
64. S. Eigler, Graphene an introduction to the fundamentals and industrial applications herausgegeben von madhuri sharon und maheshwar sharon. Angewandte Chemie **128**, (2016). https://doi.org/10.1002/ange.201602067
65. S. Yong Un Jung, T.H.S.W.C. Kyung-Won Park, S.J. Kang, High-transmittance liquid-crystal displays using graphene conducting layers. Liquid Crystals **41**(1), 101–105 (2014). https://doi.org/10.1080/02678292.2013.83751710.1080/02678292.2013.837517
66. J. Sun, C. Hu, B. Wu, H. Liu, J. Qu, Improving ion rejection of graphene oxide conductive membranes by applying electric field. J. Membr. Sci. **604**, 118077 (2020)
67. B. Li, W. Tang, Y. Zhou, J. Liu, D. Sun, X. Wang, G. Zhang, B. Li, Y. Ge, Ultrasonic-mediated electrochemical design of graphene/polyacrylonitrile conductive mem- brane for antifouling and electrofiltration. Sep. Purif. Technol. **326**, 124727 (2023)
68. S. Lin, Y. Lu, J. Xu, S. Feng, J. Li, High performance graphene/semiconductor van der waals heterostructure optoelectronic devices. Nano Energy **40**, 122–148 (2017)
69. X. Wan, Y. Huang, Y. Chen, Focusing on energy and optoelectronic applications: a journey for graphene and graphene oxide at large scale. Acc. Chem. Res. **45**(4), 598–607 (2012)
70. Q. Ke, J. Wang, Graphene-based materials for supercapacitor electrodes–a review. J Materiomics **2**(1), 37–54 (2016)
71. J. Bogus-lawski, Y. Wang, H. Xue, X. Yang, D. Mao, X. Gan, Z. Ren, J. Zhao, Q. Dai, G. Sobon´ et al, Graphene actively mode-locked lasers. Adv. Funct. Mater. **28**(28), 1801539 (2018)
72. G. Sobon, J. Sotor, Recent advances in ultrafast fiber lasers mode-locked with graphenebased saturable absorbers. Curr. Nanosci. **12**(3), 291–298 (2016)
73. Z. Liu, X. Zhang, X. Yan, Y. Chen, J. Tian, Nonlinear optical properties of graphene-based materials. Chin. Sci. Bull. **57**, 2971–2982 (2012)
74. M. Junaid, M. Md Khir, G. Witjaksono, Z. Ullah, N. Tansu, M.S.M. Saheed, P. Kumar, L. Hing Wah, S.A. Magsi, M.A. Siddiqui, A review on graphene-based light emitting functional devices. Molecules **25**(18), 4217 (2020)
75. T. Gokus, R. Nair, A. Bonetti, M. Bohmler, A. Lombardo, K. Novoselov, A. Geim, A. Ferrari, A. Hartschuh, Making graphene luminescent by oxygen plasma treatment. ACS Nano **3**(12), 3963–3968 (2009)
76. G. Eda, Y.-Y. Lin, C. Mattevi, H. Yamaguchi, H.A. Chen, I. Chen, C.W. Chen, M. Chhowalla et al., Blue photoluminescence from chemically derived graphene oxide (2009). arXiv:0909.2456
77. J. Wang, F. Ma, W. Liang, R. Wang, M. Sun, Optical, photonic and optoelectronic properties of graphene, h-bn and their hybrid materials. Nanophotonics **6**(5), 943–976 (2017)
78. M. Massicotte, G. Soavi, A. Principi, K.-J. Tielrooij, Hot carriers in graphene–fundamentals and applications. Nanoscale **13**(18), 8376–8411 (2021)
79. G. Kalita, M. Masahiro, H. Uchida, K. Wakita, M. Umeno, Few layers of graphene as transparent electrode from botanical derivative camphor. Mater. Lett. **64**(20), 2180–2183 (2010)
80. Y.U. Jung, K.-W. Park, S.-T. Hur, S.-W. Choi, S.J. Kang, High-transmittance liquid-crystal displays using graphene conducting layers. Liq. Cryst. **41**(1), 101–105 (2014)

81. R. Singh, D. Kumar, C. Tripathi, Graphene: potential material for nanoelectronics applications. NISCAIR-CSIR (2015)
82. H. Chen, G. Mi, P. Li, X. Huang, C. Cao, Microstructure and tensile properties of graphene-oxide-reinforced high-temperature titanium-alloy-matrix composites. Materials **13**(15), 3358 (2020)
83. F. D'Souza, K.M. Kadish, Handbook of carbon nano materials. World Scientific (2015)
84. C.C. Koch, I.A. Ovid'Ko, S. Seal, S. Veprek, Structural nanocrystalline materials: fundamentals and applications. Cambridge University Press (2007)
85. I. Ovid'Ko,Mechanical properties of graphene. Rev. Adv. Mater. Sci **34**(1), 1–11 (2013)
86. A.A. Griffith, Vi. the phenomena of rupture and flow in solids. Philosophical transactions of theroyal society of london. Ser. A, Contain. Pap. Math. Phys. Character **221**(582–593), 163–198 (1921)
87. F. Liu, P. Ming, J. Li, Ab initio calculation of ideal strength and phonon instability of graphene under tension. Phys. Rev. B—Condensed Matter Mater. Phys. **76**(6), 064120 (2007)
88. T. Zhu, J. Li, Ultra-strength materials. Prog. Mater Sci. **55**(7), 710–757 (2010)
89. K.N. Kudin, G.E. Scuseria, B.I. Yakobson, C 2 f, bn, and c nanoshell elasticity from ab initio computations. Phys. Rev. B **64**(23), 235406 (2001)
90. F. Banhart, J. Kotakoski, A.V. Krasheninnikov,Structural defects in graphene. ACS nano (1), 26–41 (2011)
91. T. Xu, L. Sun, Structural defects in graphene. In *Defects in Advanced Electronic Materials and Novel Low Dimensional Structures*, pp. 137–160. Elsevier (2018)
92. A. Hashimoto, K. Suenaga, A. Gloter, K. Urita, S. Iijima, Direct evidence for atomic defects in graphene layers. Nature **430**(7002), 870–873 (2004)
93. K.I. Tserpes,Strength of graphenes containing randomly dispersed vacancies. Acta Mechanica (4), 669–678 (2012)
94. H. Yin, H.J. Qi, F. Fan, T. Zhu, B. Wang, Y. Wei, Griffith criterion for brittle fracture in graphene. Nano Lett. **15**(3), 1918–1924 (2015)
95. W. Luo, K. Yarn, Z. Zheng, F. Fasya, D. Faridah, C. Chen, Performance analysis of direct methanol fuel cell with catalyst and graphene mixture coated on to fuel channels. Dig. J. Nanomater. Biostruct. **13**(3), 765–775 (2018)
96. S.-E. Zhu, S. Yuan, G. Janssen, Optical transmittance of multilayer graphene. Europhys. Lett. **108**(1), 17007 (2014)
97. E. Yamasue, M. Susa, H. Fukuyama, K. Nagata, Thermal conductivities of silicon and germanium in solid and liquid states measured by non-stationary hot wire method with silica coated probe. J. Cryst. Growth **234**(1), 121–131 (2002)
98. D. Thuau, I. Koymen, R. Cheung, A microstructure for thermal conductivity measurement of conductive thin films. Microelectron. Eng. **88**(8), 2408–2412 (2011)
99. B.K. Lee, Y.-H. Song, J.-B. Yoon, Indium tin oxide (ito) transparent mems switches, in *2009 IEEE 22nd International Conference on Micro Electro Mechanical Systems* IEEE, pp. 148–151 (2009)
100. N.M. Ravindra, S.R. Marthi, A. Banobre,Radiative Properties of Semiconductors. IOP Concise Physics. Morgan & Claypool Publishers. (2017). https://books.google.com.au/books?id=Fj9iDwAAQBAJ
101. D. Casari, L. Peth¨o, P. Schu¨rch, X. Maeder, L. Philippe, J. Michler, P. Zysset, P., Schwiedrzik, J.: A self-aligning microtensile setup: Application to single-crystal gaas microscale tension–compression asymmetry. Journal of Materials Research **34**(14), 2517–2534 (2019)
102. K.S. Novoselov, L. Colombo, P. Gellert, M. Schwab, K. Kim et al., A roadmap for graphene. nature **490**(7419), 192–200 (2012)
103. X. Wang, G. Shi, An introduction to the chemistry of graphene. Phys. Chem. Chem. Phys. **17**(43), 28484–28504 (2015)
104. S. Chen, L. Brown, M. Levendorf, W. Cai, S.-Y. Ju, J. Edgeworth, X. Li, C.W. Magnuson, A. Velamakanni, R.D. Piner et al., Oxidation resistance of graphene-coated cu and cu/ni alloy. ACS Nano **5**(2), 1321–1327 (2011)

105. L. Yan, Y.B. Zheng, F. Zhao, S. Li, X. Gao, B. Xu, P.S. Weiss, Y. Zhao, Chemistry and physics of a single atomic layer: strategies and challenges for functionalization of graphene and graphene-based materials. Chem. Soc. Rev. **41**(1), 97–114 (2012)
106. C. Huang, C. Li, G. Shi, Graphene based catalysts. Energy Environ. Sci. **5**(10), 8848–8868 (2012)
107. K.M. Yam, N. Guo, Z. Jiang, S. Li, C. Zhang, Graphene-based heterogeneous catalysis: role of graphene. Catalysts **10**(1), 53 (2020)
108. L. Kong, A. Enders, T.S. Rahman, P.A. Dowben, Molecular adsorption on graphene. J. Phys.: Condens. Matter **26**(44), 443001 (2014)
109. M.Z. Tonel, I. Zanella, S.B. Fagan, Theoretical study of small aromatic molecules adsorbed in pristine and functionalised graphene. J. Mol. Model. **27**(6), 193 (2021)
110. N. Manousi, E. Deliyanni, E. Rosenberg, G. Zachariadis, Ultrasound-assisted magnetic solid-phase extraction of polycyclic aromatic hydrocarbons and nitrated polycyclic aromatic hydrocarbons from water samples with a magnetic polyaniline modified graphene oxide nanocomposite. J. Chromatogr. A **1645**, 462104 (2021)
111. S. Debnath, Q. Cheng, T.G. Hedderman, H.J. Byrne, A raman spectroscopy study of the solubilisation of swcnts by polycyclic aromatic hydrocarbons. Carbon **48**(5), 1489–1497 (2010)
112. S. Debnath, Q. Cheng, T.G. Hedderman, H.J. Byrne, An experimental study of the interaction between single walled carbon nanotubes and polycyclic aromatic hydrocarbons. Physica status solidi (b) **245**(10), 1961–1963 (2008)
113. O.V. Ershova, T.C. Lillestolen, E. Bichoutskaia, Study of polycyclic aromatic hydrocarbons adsorbed on graphene using density functional theory with empirical dispersion correction. Phys. Chem. Chem. Phys. **12**(24), 6483–6491 (2010)
114. S. Debnath, Selective solubilisation of single walled carbon nanotubes using polycyclic aromatic hydrocarbons. PhD thesis, Technological University Dublin, Ireland (2010)
115. T. Kuila, S. Bose, A.K. Mishra, P. Khanra, N.H. Kim, J.H. Lee, Chemical functionalization of graphene and its applications. Prog. Mater Sci. **57**(7), 1061–1105 (2012)
116. G. Bottari, M.A. Herranz, L. Wibmer, M. Volland, L. Rodr´ıguez-P´erez, D.M. Guldi, A. Hirsch, N. Mart´ın, F. D'Souza, T. Torres, Chemical functionalization and characterization of graphene-based materials. Chem. Soc. Rev. **46**(15), 4464–4500 (2017)
117. E. Bekyarova, S. Sarkar, F. Wang, M.E. Itkis, I. Kalinina, X. Tian, R.C. Haddon, Effect of covalent chemistry on the electronic structure and properties of carbon nanotubes and graphene. Acc. Chem. Res. **46**(1), 65–76 (2013)
118. P.A. Denis, F. Iribarne, Comparative study of defect reactivity in graphene. J. Phys. Chem. C **117**(37), 19048–19055 (2013)
119. G. Zhao, X. Li, M. Huang, Z. Zhen, Y. Zhong, Q. Chen, X. Zhao, Y. He, R. Hu, T. Yang et al., The physics and chemistry of graphene-on surfaces. Chem. Soc. Rev. **46**(15), 4417–4449 (2017)
120. J.E. Johns, M.C. Hersam, Atomic covalent functionalization of graphene. Acc. Chem. Res. **46**(1), 77–86 (2013)
121. B. Nazari, Z. Ranjbar, R.R. Hashjin, A.R. Moghaddam, G. Momen, B. Ranjbar, Dispers- ing graphene in aqueous media: investigating the effect of different surfactants. Colloids Surf., A **582**, 123870 (2019)
122. S. Qamar, N. Ramzan, W. Aleem, Graphene dispersion, functionalization techniques and applications: a review. Synth. Metals 117697 (2024)
123. J. Park, M. Yan, Covalent functionalization of graphene with reactive intermediates. Acc. Chem. Res. **46**(1), 181–189 (2013)
124. L.J. Cote, F. Kim, J. Huang, Langmuir-blodgett assembly of graphite oxide single layers. J. Am. Chem. Soc. **131**(3), 1043–1049 (2009)
125. J. Kim, L.J. Cote, F. Kim, W. Yuan, K.R. Shull, J. Huang, Graphene oxide sheets at interfaces. J. Am. Chem. Soc. **132**(23), 8180–8186 (2010)
126. D.R. Dreyer, A.D. Todd, C.W. Bielawski, Harnessing the chemistry of graphene oxide. Chem. Soc. Rev. **43**(15), 5288–5301 (2014)

127. S. Zhou, A. Bongiorno, Origin of the chemical and kinetic stability of graphene oxide. Sci. Rep. **3**(1), 2484 (2013)
128. R. Tarcan, O. Todor-Boer, I. Petrovai, C. Leordean, S. Astilean, I. Botiz, Reduced graphene oxide today. J. Mater. Chem. C **8**(4), 1198–1224 (2020)
129. S. Gupta, P. Joshi, J. Narayan, Electron mobility modulation in graphene oxide by controlling carbon melt lifetime. Carbon **170**, 327–337 (2020)
130. A. Bhaumik, A. Haque, M. Taufique, P. Karnati, R. Patel, M. Nath, K. Ghosh, Reduced graphene oxide thin films with very large charge carrier mobility using pulsed laser deposition. J. Mater. Sci. Eng. **6**(4), 1–11 (2017)
131. J. Chen, L. Li, Thermal conductivity of graphene oxide: a molecular dynamics study. JETP Lett. **112**, 117–121 (2020)
132. P. Kumar, F. Shahzad, S. Yu, S.M. Hong, Y.-H. Kim, C.M. Koo, Large-area reduced graphene oxide thin film with excellent thermal conductivity and electromagnetic interference shielding effectiveness. Carbon **94**, 494–500 (2015)
133. A. Benchirouf, C. Mu¨ller, O. Kanoun, Electromechanical behavior of chemically reduced graphene oxide and multi-walled carbon nanotube hybrid material. Nanoscale Res. Lett. **11**, 1–7 (2016)
134. A. Teklu, C. Barry, M. Palumbo, C. Weiwadel, N. Kuthirummal, J. Flagg, Mechanical characterization of reduced graphene oxide using afm. Adv. Condens. Matter Phys. **2019**(1), 8713965 (2019)
135. P. Montes-Navajas,N.G. Asenjo, R. Santamar´ıa, R. Men´endez, A. Corma, H. Garc´ıa, Surface area measurement of graphene oxide in aqueous solutions. Langmuir **29**(44), 13443–13448 (2013)
136. H.-B. Zhang, J.-W. Wang, Q. Yan, W.-G. Zheng, C. Chen, Z.-Z. Yu, Vacuum-assisted synthesis of graphene from thermal exfoliation and reduction of graphite oxide. J. Mater. Chem. **21**(14), 5392–5397 (2011)
137. Z. Qiao, C. Qin, Y. Gao, G. Zhang, R. Chen, L. Xiao, S. Jia, Modulation of the optical transmittance in monolayer graphene oxide by using external electric field. Sci. Rep. **5**(1), 14441 (2015)
138. J. Wan, F. Gu, W. Bao, J. Dai, F. Shen, W. Luo, X. Han, D. Urban, L. Hu, Sodium- ion intercalated transparent conductors with printed reduced graphene oxide networks. Nano Lett. **15**(6), 3763–3769 (2015)
139. D. Konios, M.M. Stylianakis, E. Stratakis, E. Kymakis, Dispersion behaviour of graphene oxide and reduced graphene oxide. J. Colloid Interface Sci. **430**, 108–112 (2014)
140. W. Zhang, L. Wu, Z. Li, Y. Liu, Doped graphene: synthesis, properties and bioanalysis. RSC Adv. **5**(61), 49521–49533 (2015)
141. R. Kumar, S. Sahoo, E. Joanni, R.K. Singh, K. Maegawa, W.K. Tan, G. Kawamura, K.K. Kar, A. Matsuda, Heteroatom doped graphene engineering for energy storage and conversion. Mater. Today **39**, 47–65 (2020)
142. X. Wang, G. Sun, P. Routh, D.-H. Kim, W. Huang, P. Chen, Heteroatom-doped graphene materials: syntheses, properties and applications. Chem. Soc. Rev. **43**(20), 7067–7098 (2014)
143. W. Xiluan, G. Shi, Introduction to the chemistry of graphene. Phys. Chem. Chem. Phys. PCCP **17** (2015). https://doi.org/10.1039/c5cp05212b
144. C.N.R. Rao, K. Gopalakrishnan, A. Govindaraj, Synthesis, properties and applications of graphene doped with boron, nitrogen and other elements. Nano Today **9**(3), 324–343 (2014)
145. Q. Li, J. Zhu, Y. Hou, S. Sun, Graphene and its composites with nanoparticles for electrochemical energy applications. Nano Today (2014). https://doi.org/10.1016/j.nantod.201409.002
146. Y. Wen, C. Huang, L. Wang, D. Hulicova-Jurcakova, Heteroatom-doped graphene for electrochemical energy storage. Chin. Sci. Bull. **59**, 2102–2121 (2014)
147. B. Yao, C. Li, J. Ma, G. Shi, Porphyrin-based graphene oxide frameworks with ultra-larged-spacings for the electrocatalyzation of oxygen reduction reaction. Phys. Chem. Chem. Phys. **17**(29), 19538–19545 (2015)

148. Z. Peralta-Inga, J. Murray, M.E. Grice, S. Boyd, C. O'Connor, P. Politzer, Computa- tional characterization of surfaces of model graphene systems. J. Mol. Struct. (Thoechem) **549**(1–2), 147–158 (2001)
149. R. Sharma, N. Nair, M.S. Strano, Structure- reactivity relationships for graphene nanoribbons. J. Phys. Chem. C **113**(33), 14771–14777 (2009)
150. A. Eftekhari, H. Garcia, The necessity of structural irregularities for the chemical applications of graphene. Mater. Today Chem. **4**, 1–16 (2017)
151. A. Bellunato, H. Arjmandi Tash, Y. Cesa, G.F. Schneider, Chemistry at the edge of graphene. ChemPhysChem **17**(6), 785–801 (2016)
152. C. Hu, D. Liu, Y. Xiao, L. Dai, Functionalization of graphene materials by heteroatom- doping for energy conversion and storage. Prog. Nat. Sci.: Mater. Int. **28**(2), 121–132 (2018)
153. S.D. Costa, J.E. Weis, O. Frank, M. Kalbac, Effect of layer number and layer stacking registry on the formation and quantification of defects in graphene. Carbon **98**, 592–598 (2016)
154. Y. Ding, R. Wu, I.H. Abidi, H. Wong, Z. Liu, M. Zhuang, L.-Y. Gan, Z. Luo, Stacking modes-induced chemical reactivity differences on chemical vapor deposition-grown trilayer graphene. ACS Appl. Mater. Interfaces. **10**(27), 23424–23431 (2018)
155. R. Sharma, J.H. Baik, C.J. Perera, M.S. Strano, Anomalously large reactivity of single graphene layers and edges toward electron transfer chemistries. Nano Lett. **10**(2), 398–405 (2010)

Chapter 2
Production of Graphene

Abstract This chapter offers a critical examination of current graphene production techniques, drawing on an extensive review of the literature to evaluate both bottom-up and top-down approaches. Bottom-up methods, such as chemical vapour deposition (CVD) and epitaxial growth, construct graphene from molecular precursors and are known for producing high-quality, low-defect material. However, these techniques often involve costly substrates, sophisticated equipment, and significant material waste—factors that hinder scalability and broader commercial uptake. Conversely, top-down methods, including mechanical, liquid-phase, and electrochemical exfoliation of graphite, offer more accessible and scalable routes to graphene synthesis. Liquid-phase exfoliation has shown promise for low-cost, bulk production, although challenges remain in controlling flake size, layer number, and defect density. The chapter underscores the need for continued innovation to reconcile the trade-off between material quality and process scalability. By identifying key gaps in current methodologies and proposing potential research directions, this analysis aims to support the development of more efficient, sustainable, and application-oriented graphene fabrication techniques. It contributes to the foundational understanding necessary for advancing graphene-based technologies across materials science, nanotechnology, and engineering.

Keywords Graphene · Production methods · Bottom-up approach · Top-down approach · Chemical vapour deposition (CVD) · Epitaxial growth · Liquid-phase exfoliation · Mechanical exfoliation · Electrochemical exfoliation · Graphene's derivatives

2.1 Introduction

Graphene exhibits great potential as a versatile material for a wide range of applications, including electronics, biomedical engineering, energy storage, and numerous other fields. Although numerous scientific research progresses have been reported

S. Debnath et al., *Graphene in Wearable Sensors for Health Monitoring*, SpringerBriefs in Applied Sciences and Technology,
https://doi.org/10.1007/978-981-96-8850-0_2

regarding the application of graphene and graphene-based material for wide and practical applications, as well as commercialisation, the cost-effective, scalable production, and commercial availability of graphene are essential. Various methods have been proposed for graphene production thus far, broadly categorised into two main approaches: bottom-up and top-down [1].

The bottom-up approach uses chemical reactions involving molecular building blocks to create covalently linked two-dimensional networks. Bottom-up techniques, such as chemical vapour deposition (CVD) and epitaxial growth (EG), produce high-quality graphene with minimal defects and are well-suited for electronic applications [2–4]. However, substrate-based methods require expensive equipment, complex process control, and high material wastage, etc. This reduces scalability and increases production costs, making it challenging to meet the demand for large quantities of graphene. In contrast, the top-down method involves exfoliating graphite, which separates the layers of graphite to produce graphene. The exfoliation of graphite can occur through mechanical means, which affect the application of forces that generate either longitudinal or transverse stress on the surface of the graphite [5]. Furthermore, liquid-phase exfoliation and chemical and electrochemical exfoliation techniques may be employed by introducing large alkali ions into the interlayer space of graphite within a dispersed solution. The top-down methods have shown that large-scale, low-cost graphene production is feasible through the direct exfoliation of graphite in a liquid medium [6–8].

The following sections discuss different bottom-up and top-down approaches for graphene production.

2.2 The Bottom-Up Approach

2.2.1 Epitaxial Growth (EG)

Epitaxial growth refers to depositing a monocrystalline film onto a monocrystalline substrate, resulting in an epitaxial film that maintains a consistent atomic arrangement throughout the layers [9]. EG produces graphene as a thin surface layer. This method has been proposed as a promising alternative to silicon as a semiconducting material for microchips [10, 11] and as a biosensor for detecting cancer and other diseases [12]. EG is generally formed through the thermal synthesis of graphene films on a single-crystal silicon carbide (SiC) surface [9, 13]. With its high quality and direct growth on SiC semiconductor substrates, this method is particularly suitable for these applications. This ensures high quality and saves time and resources by eliminating the need for a transfer process [14, 15].

In 2004, Berger et al. successfully fabricated ultrathin epitaxial graphite films, with thicknesses comprising only a few layers, on the Si-terminated plane of single-crystal 6H-SiC through the thermal desorption of silicon. The samples underwent hydrogen etching and were subjected to electron bombardment heating in an ultrahigh vacuum to eliminate any existing oxide layers. Typically, multilayered EG films comprised of three graphene sheets with lateral dimensions of several tens of nanometers were synthesised, with temperature being the primary determinant of layer thickness [16].

A subsequent study by Rollings et al. in 2006 focused on synthesising graphene films on the Si-terminated face of n-type 6H-SiC wafers. These samples were placed within an ultrahigh-vacuum preparation chamber and annealed at 850 °C under silicon flux for 20 to 30 min to remove surface contaminants. This process yielded EG graphene with thicknesses ranging from 1 to 2 layers, with some samples achieving dimensions in the micrometre range [17]. In the same year, Berger et al. produced exceptionally thin EG graphene on both Si-terminated and C-terminated faces of 4H-SiC, realising films with lateral sizes measuring a few micrometres and exhibiting structural coherence exceeding 90 nm [18].

Xu et al. reported the rapid growth of a graphene film measuring (5 × 50) cm^2 in just 20 min, boasting over 99% highly-oriented grains in 2017. This achievement was made possible through several steps, such as the fabrication of a metre-sized single-crystal Cu(1 1 1) foil as the substrate, the epitaxial growth of graphene islands on the Cu(1 1 1) surface, the seamless integration of these graphene islands into a cohesive film with high single crystallinity, and the fast growth of the graphene film itself. The successful outcome was facilitated by a temperature-gradient-driven annealing technique that transformed industrial polycrystalline copper foil into single-crystal Cu(1 1 1) alongside the beneficial effects of a continuous oxygen supply from a nearby oxide source. The resulting graphene film, which features very few misoriented grains, exhibits a mobility of up to approximately 23,000 $cm^2 V^{-1} s^{-1}$ at 4 K and a room temperature sheet resistance of around 230 Ωcm^{-2}[19].

2.2.2 *Chemical Vapour Deposition (CVD)*

Chemical vapour deposition (CVD) has emerged as a cost-effective technique for producing high-quality monolayer graphene, large-area graphene on transition metal substrates, which is suitable for many applications [20]. Both large-area poly-crystalline and millimetre-sized single-crystal graphene domains were successfully synthesised using CVD [21–24]. Large-area graphene samples can be generated by exposing metallic substrates to various hydrocarbon precursors at high temperatures [22, 25, 26].

To synthesise the desired product effectively, a particular configuration of a tube-furnace CVD system is necessary, typically comprising a gas delivery system, a reactor, and a gas removal system. In the context of graphene production, the process generally begins with the introduction of a precise mixture of gases into the chamber, which includes a carbon precursor (such as methane (CH_4)), gas carriers (like nickel or argon), and occasionally hydrogen (H_2) as an activating agent. The precursor was then decomposed by heat, allowing it to react with the substrate and form graphene on its surface, typically on copper (Cu) foils or films. The process concludes with the expulsion of by-products and unreacted gases during cooling [27, 28]. As the substrate cools naturally, the carbon diffuses from the substrate and precipitates on its surface due to its low solubility in metals at lower temperatures. This process results in the formation of mono- to multilayer graphene sheets on the substrate surface. Notably, graphene synthesised on copper foils is characterised by high uniformity and superior quality compared to that produced on other transition metals, presumably due to copper's lower affinity for carbon in contrast to nickel and cobalt [3, 29].

In 2006, Somani et al. utilised camphor as the precursor material and nickel substrates for carbon deposition, ultimately producing 'planar few-layer graphene', which consists of approximately 35 graphitic layers. This effort encouraged further exploration and refinement of the process [30]. Nickel typically yields multilayer graphene due to its high solubility for carbon. In contrast, the initial synthesis of graphene on copper was first demonstrated by Li et al. through the utilisation of a methane and hydrogen mixture at 1000 °C in a low-pressure quartz tube reactor [31].

The inherent low carbon solubility in copper, combined with the limited carbon saturation caused by graphene film coverage, results in predominantly single-layer graphene with coverage exceeding 95%. Regarding scalability, CVD is a favourable method for manufacturing high-quality graphene in larger quantities, prompting numerous research groups to focus on this area. A notable early effort by Bae et al. [32] employed a roll-to-roll (RTR) technique to manufacture 30-inch graphene films, pushing forward continuous graphene production via this approach [33–35]. Similarly, Polsen et al. integrated the RTR process with concentric tube CVD for enhanced graphene manufacturing processes [36]. Lin et al. introduced an innovative surface engineering approach, successfully achieving centimetre-sized single-crystalline graphene by passivating active sites and regulating nucleation through melamine pretreatment of the copper substrate [37]. Furthermore, Bointon et al. made considerable progress using resistive-heating cold-wall CVD to produce high-quality monolayer graphene at a speed 100 times greater than conventional CVD [38]. While these studies show promising advancements, the mass production of graphene via CVD has mainly targeted the electronics sector.

2.2.2.1 Plasma-Enhanced CVD (PECVD)

In PECVD, high-energy electrons produced by a plasma generator lead to the ionisation, excitation, and dissociation of hydrocarbon precursors at lower temperatures than traditional CVD methods. The PECVD setup consists of three main components: the gas, the plasma generator, and the vacuum heating chamber. Plasma generators can be divided into three categories based on their power source: microwave (MW) plasma (2.45 GHz), radio frequency (RF) plasma (13.56 MHz), and direct current (DC) plasma [22, 39]. Consequently, graphene can be directly grown on desired substrates at low temperatures using PECVD without the need for metal catalysts. As a result, PECVD has gained increasing attention as a promising method for the controlled synthesis of graphene [22].

2.2.2.2 Hot and Cold Wall CVD

In standard hot-wall CVD (HW-CVD) systems, such as quartz tube furnaces, the reactor is heated from the outside, allowing the energy to decompose carbon sources and facilitate graphene growth. This heating method produces a uniform distribution of high temperatures throughout the reaction chamber, as evidenced by computational fluid dynamics (CFD) simulations [40]. Consequently, the gas phase temperature over the growth substrate remains sufficiently high to trigger side reactions, which produce various carbon species that contribute to the formation of amorphous carbon. Also, hot-wall CVD reactors require long ramping-up and cooling-down procedures, which are time-consuming and limit the throughput for industrial-scale synthesis of graphene [38, 41].

Unlike other CVD reactors, cold wall CVD (CW-CVD) reactors have several unique features, e.g. rapid sample heating and cooling, which lead to faster growth rates [38, 41, 42]. Furthermore, the reduced chemical reactions in the gas phase of the CW-CVD reactors provide unparalleled control over film quality. These reactors also exhibit lower heat capacities by employing localised heating adjacent to the sample, resulting in decreased power consumption. As a result, these reactors enable accelerated growth rates and superior film quality control, operating with impressive energy efficiency. They are an optimal system for producing super-clean graphene films, as their unique heat distribution suppresses gas-phase reactions by lowering the gas-phase temperature [40].

Generally, the processes that regulate graphene formation, e.g. PECVD, HW-CVD, and CW-CVD, are influenced by the selection of the growth substrate. The standard procedure involves the nucleation of carbon atoms on the metallic substrate following the thermal decomposition of hydrocarbons. This results in the growth of larger graphene domains. Aside from gaseous hydrocarbons such as methane, ethylene, and acetylene, liquid precursors, including hexane and pentane, have also demonstrated their ability in synthesising graphene [31]. However, CVD techniques pose some challenges. For instance, transferring graphene from the growth substrate to a target substrate can be challenging due to its chemical inertness, which may result in defects and wrinkles [43].

Nevertheless, for electronic applications, CVD graphene requires a transfer process post-growth from the catalyst surface to dielectric substrates, which can introduce contamination and structural defects, thereby diminishing the performance of graphene. Graphene's growth requires higher temperatures in the absence of a metal catalyst. Several studies reported the direct growth of graphene on dielectric substrates, requiring elevated temperatures ranging from 1100 to 1650 °C [44–46]. Additionally, thermal fluctuations may influence the stability of the produced graphene and the complexity of the CVD process and the high energy requirements for the specific method increase the difficulties of the production. Despite these challenges, CVD remains one of the most effective techniques for producing large-area graphene [47].

2.3 The Top-Down Approach

2.3.1 *Exfoliation*

Graphite consists of layers of graphene sheets held together by van der Waals forces. Consequently, graphene can be derived from graphite by breaking these connections. In the top-down approach to graphene production, graphite is exfoliated to isolate graphene layers. This can be accomplished through mechanical or chemical exfoliation, which involves repeated peeling [48].

Mechanical exfoliation is the most direct method for producing graphene. It is distinguished as the technique that led to the creation of standalone graphene. The ideal scenario involves peeling graphene off from the bulk graphite layer by layer. A fundamental mechanical challenge is overcoming the Van der Waals forces that hold neighbouring graphene flakes together. It is reported that the inter-layer van der Waals interaction energy of approximately 2 eV $\cdot$ nm^{-2} suggests that the force required to exfoliate graphite is around 300 nN μm^{-2} [49]. This small force can be easily achieved using adhesive tape. This approach has proven reliable and was first introduced in 2004 by Novoselov and Geim [50], for which they received the Nobel Prize in 2010 [51, 52]. They employed Scotch Tape to repeatedly peel layers from the bulk of highly ordered pyrolytic graphite (HOPG), successfully isolating graphene sheets from HOPG flakes. This method statistically reduces a 1 μm thick graphite flake to a monolayer-thin sample [53]. The smooth, thin fragments deposited onto the tape were first scrutinised via optical microscopy to evaluate their integrity and uniformity before carefully transferring them onto a cleaned substrate. Selecting an appropriate substrate was an important factor in the experimental design. Their work presented a straightforward method for graphene isolation and demonstrated that optical identification of the material could be achieved by placing it on appropriately thick Si/SiO_2 substrates [50]. The contrast of a graphene monolayer on a Si/SiO_2 substrate, featuring an oxide layer of either 300 or 90 nm, was optimised to roughly 12% at a wavelength of 550 nm, aligning with the peak sensitivity of human vision. Consequently, thicker HOPG flakes placed on a 300 nm SiO_2 substrate appear yellow to bluish. At the same time, as thickness decreases to sub-10 nm levels, they display shades from darker to lighter purple, indicating multiple layers, ultimately revealing a single monolayer of graphene. The visibility of graphene arises from the interference effects in the substrate, allowing for modifications to the substrate by adjusting the thickness of the dielectric layer as long as some reflectivity is preserved. A disadvantage of the tape method is that it may leave adhesive residues on the sample. Such residues were reported to impede carrier mobility [54, 55]. Therefore, heat treatment after deposition, potentially in a reducing atmosphere, can be applied to eliminate the residue [56, 57]. Moreover, although this method yields the highest quality crystals, it is highly labour-intensive and time-consuming, and is confined to laboratory-scale experiments and prototyping, as it cannot be scaled up efficiently [1, 52].

2.3.2 *Ball Milling*

Ball milling is a widely used mechanical exfoliation. Exfoliation occurs through shear force and the collisions caused by the balls during rolling, which leads to the fragmentation of graphite flakes and can even damage their crystalline structure. There are two categories of ball milling: wet ball milling, which involves solvents, and dry ball milling, which utilises chemically inert, water-soluble inorganic salts [1]. Initially, ball milling was used to reduce graphite size, producing graphitic flakes as thin as 10 nm [58, 59]. However, this milling approach had yet to be optimised for creating graphene.

2.3.2.1 Wet Ball Milling

In 2010, Zhao's research group reported the successful exfoliation of graphite platelets to yield crystalline graphene sheets through a wet ball milling process conducted in a liquid medium. The team systematically dispersed 0.02 g of multilayered graphite nanosheets (GNs) in 80 ml of anhydrous N, N-dimethylformamide (DMF) at a concentration of 0.25 $mg.ml^{-1}$. They performed shear force-dominated ball milling for 30 h in a planetary mill utilising zirconia balls and poly(tetrafluoroethylene) vials, maintaining a low rotation speed of 300 rpm to preserve the shear stress regime. Post-milling, the dispersion was subjected to centrifugation to remove larger particulate matter. The resulting cross-sectional thickness of the graphene sheets was measured to range between 0.8 and 1.8 nm, indicative of single-layer and few-layer graphene structures, with an approximate thickness corresponding to three layers [1]. Knieke et al. [60] and Damm et al. [61] employed wet-stirred media mills that operate with smaller grinding media. This process allows better temperature control during processing. They optimised technical aspects, such as the milling tool, the size of the delamination media, and the rotation speed of the stirrer. Notably, using a shaking plate as the milling apparatus has been reported to increase the concentration of dispersed carbon, with optimal results observed at a ZrO_2 bead size of 100 μm, yielding a maximal concentration of few-layer graphene (FLG).

2.3.2.2 Dry Ball Milling

In addition to the established wet milling method, dry milling has emerged as a viable technique for producing graphene. The process involves altering the layers of graphite through the ball milling of a mixture containing graphite and chemically inert, water-soluble inorganic salts. After this process, washing the milled product with water or using sonication can yield graphene powders [62, 63]. The advantages of dry milling are particularly pronounced when combined with functionalization and exfoliation strategies. For instance, an investigation into dry ball milling techniques was conducted by Lin et al. in 2013, who employed a shear-force-dominated grinding technique with a ball mill using elemental sulphur (S_8) to promote the exfoliation adhering to graphene sheets, resulting in a composition consisting of 73 wt.% sulphur [64]. Sulphur is a moderately adhesive material for the graphite plane, functioning similarly to Scotch tape, allowing the team to create highly crystalline graphene films. The electronegativities of sulphur and graphene establish strong interactions, producing graphene/sulphur composites by ball-milling chemically modified graphite with elemental sulphur.

An Atomic Force Microscopy (AFM) instrument's tip also demonstrated the ability to slide graphene sheets away from graphite and even fold a single graphene sheet over itself, generating superlattices of bilayer graphene. In a parallel investigation, Leon et al. employed ball milling to exfoliate graphite, allowing for multi-point interactions with commercially available melamine under solid-state conditions facilitated by a hydrogen-bonding network [65].

Although the ball milling technique is promising for large-scale graphene production, the high-energy impacts from the grinding media can lead to defects in the graphene. Since collisions between grinding media are inevitable during milling, fragmentation and defect generation are inherent, presenting a dual challenge.

2.3.3 *Liquid Phase Exfoliation (LPE)*

Liquid-phase exfoliation (LPE) is one of the most promising methods for scaling up graphene production. The ease, speed, and high throughput associated with this technique make it appealing for large-scale graphene manufacturing. LPE could produce graphene of 100 L with a maximum concentration of less than 0.1 mg mL^{-1} [66]. Three fundamental stages characterise it: (1) the dispersion of graphite in a suitable solvent or surfactant, (2) the exfoliation of the material, and (3) the purification of the resultant product to effectively separate exfoliated graphene from unexfoliated materials and to remove residual solvents, particularly when utilising powdered graphite as the starting material [67].

A notable development was reported by Paton et al., who demonstrated that high shear forces could be employed as an alternative to ultrasonic cavitation, enabling the production of exfoliated graphene on a scale of 100 L [66]. They found that the critical shear rate necessary for graphene exfoliation was approximately 104 s^{-1}, which can be achieved with standard household blenders. After centrifugation, the average number of layers was fewer than 10, and the typical lateral dimensions of the nanosheets ranged from 300 to 800 nm. The yield was relatively low, and factors such as the selection of initial material and rotor optimisation can significantly influence exfoliation efficiency. At room temperature, Dimiev et al. produced graphene nanoplatelets (GNPs) for 3 to 4 h, achieving nearly a 100% conversion yield from GNPs [68].

Given the existing industrial expertise and readily available equipment, LPE currently constitutes the most practical approach for scaling graphene production

[7]. However, developing a standardised and reproducible protocol anchored in clearly defined process parameters is imperative to facilitate large-scale industrial applications by ensuring process consistency and viability.

Sonication-assisted LPE of graphite has enabled the scalable production of graphene. This method utilises ultrasonic waves to overcome the van der Waals interactions between graphite layers, thereby promoting efficient exfoliation and generating graphene materials in large quantities. However, the sonication duration is a critical parameter, as prolonged exposure is correlated with increased graphene concentration [69] and severely reduces the size of flakes, but at the expense of high energy consumption. Following the sonication process, the resultant mixture typically contains thicker graphite-like flakes, which may be eliminated through ultracentrifugation. Enhanced centrifugation speeds facilitate the attainment of thinner flakes with reduced lateral dimensions; however, these may not be optimal for specific applications, such as composite materials [7].

In 2009, Bourlinos et al. explored liquid-phase exfoliation by applying sonication in an ultrasound bath with fine graphite powder. This graphite was suspended and sonicated for one hour in perfluorinated aromatic solvents, including hexafluorobenzene (C_6F_6), octafluorotoluene ($C_6F_5CF_3$), pentafluorobenzonitrile (C_6F_5CN), and pentafluoropyridine (C_5F_5N). The resulting thick crystalline graphene sheets exhibited lateral dimensions ranging from 0.1 to 5 mm and displayed multilayered structures near the edges, although individual graphene sheets were also observed [70]. The selection of solvents is crucial for LPE, as the efficiency of graphite exfoliation and the thermodynamic stability of graphene sheets depend on appropriate solubilisation parameters [67]. A range of solvents may be employed for graphene dispersion, including aqueous surfactants [52].

Despite numerous endeavours to enhance the exfoliation process by utilising graphite [71, 72] or graphite intercalation compounds [73, 74], a challenge remains the inherently low dispersibility of pristine graphene, which behaves as a recalcitrant macromolecule. Furthermore, the most effective solvents for dissolving pristine graphene possess high boiling points, such as N-methyl-2-pyrrolidone (NMP) [7] or exhibit highly corrosive and unstable properties in atmospheric conditions, for example, chlorosulfonic acid [75]. These factors hinder subsequent processing phases. Considerable efforts have been undertaken to increase the concentration of graphene solutions via prolonged sonication [69] in various solvents and to disperse pristine graphene in solvents with lower boiling points [76] or even water by employing surfactants or polymeric agents [8]. Extended sonication durations may yield smaller graphene flakes, potentially compromising the objective of obtaining intact graphene sheets. Incorporating surfactants may escalate costs during scaling, given that only a limited subset of surfactants can attain a colloidal solution that maintains a desirable graphene concentration [77]. Furthermore, many surfactants exhibit insulating properties, necessitating an additional washing step after film formation or during device processing [8].

2.3.4 Chemical Exfoliation

In the chemical exfoliation technique, the transformation of graphite into graphene is facilitated by creating graphitic intercalation compounds via a colloidal solution [9]. This method can be divided into two primary stages. The first stage involves reducing the interlayer van der Waals forces within the stacked layers of graphite. This results in an increment within the interlayer spacing and creates a gap between the layers, which ultimately forms graphene-intercalated compounds. The subsequent phase encompasses the rapid heating or sonication of the solid graphite dispersed in a solvent. This step aims to exfoliate the graphite into a single or few graphene layers. Exfoliation may occur through chemical modifications that separate the graphite layers or by the physical intercalation of small molecules into the interlayer spaces. Large alkali ions can be inserted between the graphite layers during the treatment in solution to enhance the exfoliation process [5, 78, 79]. Viculis et al. sonicated high-purity graphite in a mixture of ethanol and KC_8 as the intercalation compound in 2003. This research highlights the potential of chemical exfoliation as a viable method for producing graphene and other carbon-based nanomaterials [80].

2.3.5 Electrochemical Exfoliation

Electrochemical exfoliation represents a specific form of chemical exfoliation that employs electrodes constructed from graphitic carbon materials, including graphite rods, carbon sheets, or highly oriented pyrolytic graphite. The method involves decomposing a graphite electrode using a liquid electrolyte and an electrical current. This process utilises electrolytes such as ionic liquids and poly (sodium-4-styrene sulfonate) in aqueous solutions. The technique involves a liquid conductive electrolyte paired with a direct current, facilitating structural expansion within graphitic materials. When a graphite electrode is subjected to an electric field, it can become intercalated either positively or negatively, leading to exfoliation as gases are released from the electrolysis of the solvent or through post-treatment [81]. The electrochemical exfoliation process yields two primary effects. The bulk graphite undergoes macroscopic expansion due to intercalation and decomposition driven by the electrolyte, which generates gas and leads to a remarkably enlarged structure. Additionally, the exfoliated materials detach from the bulk and disperse within the solution due to the release of gas. This process can occur through either anodic oxidation or cathodic reactions of the graphite-based electrode. Cathodic reaction techniques are particularly effective for producing high-quality, few-layer conductive graphene, which is well-suited for energy and optical applications [82]. The materials generated through anodic oxidation typically consist of multiple graphene layers, yield less product and exhibit oxidation states similar to those of GrO, as opposed to pure monolayer graphene [83]. The anodic oxidation of water generates graphite-intercalation compounds, promotes hydroxylation or oxidation at the edge planes of the graphite, enables the intercalation of ionic liquid anions between the graphite layers, and triggers oxidative cleavage, resulting in the precipitation of these compounds [84].

Various graphene materials can be produced depending on factors such as the applied voltage [85] and the type of electrolyte used. The range includes oxidised nanographite[86], carbon nanoribbons, nanoparticles [87], and low-defect graphene [88] derived from electrolytes, including aqueous inorganic salt solutions [89], sulfuric acid [88], lithium perchlorate in propylene carbonate [90], and various ionic liquids [91, 92]. Su et al. produced low-defect graphene through rapid electrochemical exfoliation in an economically viable sulfuric acid electrolyte [93]. In a contrasting approach, the application of a non-oxidative cathodic bias to the graphite electrode, followed by in situ electrochemical functionalisation, resulted in the generation of highly functionalised graphene, with the potential for conversion back to near-pristine graphene through the process of laser ablation [94].

One of the key benefits of electrochemical exfoliation is that it simplifies the procedure to a single step and reduces processing time, unlike many alternative methods that require more extended periods to prepare and stabilise the final product. The lateral dimensions of the resulting flakes are crucial for nanocomposite applications and are influenced by the graphite source and the conditions during the intercalation-exfoliation process. When non-oxidative salts are utilised as intercalation products, flakes with lateral sizes of up to 50 μm and thicknesses of 2 to 3 layers can be achieved [95]. Additionally, utilising liquid electrolytes or aqueous surfactants contributes to a more environmentally friendly process. Notably, employing lithium perchlorate ($LiClO_4$) as an electrolyte can yield a product that resembles GrO while avoiding harmful and toxic chemicals typically involved in conventional GrO production methods [96].

Although the advantages of using the electrochemical exfoliation method are mentioned above, several drawbacks should also be taken into consideration. A limitation of the electrochemical route lies in maintaining a continuous voltage bias on the graphite flakes throughout the exfoliation process. Most existing methodologies predominantly utilise a singular, uninterrupted piece of graphite in rods, foils, or highly oriented pyrolytic graphite slabs [8]. Electrochemical exfoliation should selectively target only the outermost graphite flakes, enabling the individual peeling off of graphene layers.

2.4 Derivatives of Graphene

Graphene has two primary derivatives: GrO and rGrO [97]. GrO is produced through the oxidation of graphite, while rGrO is created by reducing GrO. A range of functional groups, such as hydroxyl, carboxyl, and epoxy, characterise GrO. Due to its diverse functional groups, GrO can interact with various biomolecules. This makes it a strong candidate for numerous biological applications. GrO interacts effectively with other molecules—important with its enhanced strength and apparent biological activity—and is a significant material for areas with both durability and biological compatibility [98–100]. By contrast, rGrO retains some oxygen-containing groups and defects, which give it a structure similar to pristine graphene, with the amount of oxygen and defects adjustable based on the reduction process. Despite their similar 2D structures, graphene and its derivatives exhibit slight differences that result in variations in their physicochemical properties [101].

2.4.1 Graphene Oxide (GrO)

GrO is an important precursor for graphene's cost-effective large-scale production. It is synthesised via the chemical oxidation of graphite that employs concentrated acids with potent oxidising agents. This method yields a heterogeneous mixture comprising various oxygenated carbon species, along with associated by-products. GrO plays a crucial role in graphene chemistry as it represents a form of graphene modified with oxygen-containing groups. Oxidising graphene results in GrO forming, which displays a unique atomic and electronic structure characterised by small clusters of sp^2 carbon isolated within an sp^3 matrix [102, 103].

Researchers proposed diverse models to explain their frameworks [104, 105], which were based on lattice configurations. Numerous structural models were proposed for the clear explanation of the characteristics of GrO, including those formulated by Hofmann, Ruess, Scholz-Boehm, Nakajima-Matsuo, Lerf-Klinowski, and Szabo [106, 107]. Among these, the Lerf-Klinowski model is particularly notable due to its comprehensive examination using solid-state nuclear magnetic resonance and X-ray diffraction techniques [108]. Typically, GrO is synthesised through the oxidative exfoliation of graphite, followed by its reduction to graphene sheets or rGrO. Four primary methods for GrO synthesis exist Brodie, Staudenmaier, Hofmann, and Hummers [106]. Maintaining a relatively low synthesis temperature helps to keep production costs down; however, these techniques often release toxic gases like nitrogen dioxide (NO_2) and dinitrogen tetroxide (N_2O_4) [109, 110]. Therefore, evaluating process safety and environmental impact during scale-up is crucial.

Brodie first documented the initial oxidation of graphite to form GrO in 1859, who employed a combination of potassium chlorate ($KClO_3$) and fuming nitric acid ($HNO_3 + NO_2$) functioning as an intercalation agent to facilitate the expansion of graphite [111, 112]. This method was refined by Staudenmaier in 1898 [113], who utilised a mixture of nitric and sulphuric acids in a 1:4 volume ratio. The solution was cooled in an ice bath under continuous stirring, whilst graphite (4.37 g per 100 ml) was gradually introduced to mitigate exothermic reactions. The oxidising agent, potassium chlorate, was used in a weight ratio of graphite to oxidiser of 1:11. While Staudenmaier's approach is deemed safer than Brodie's, it still produces toxic gases, leading to the preference for Hummers' method. Hofmann and K¨onig further improved Brodie's method in 1937 [114]. In 1958, Hummer and Offeman introduced a novel technique for preparing GrO utilising $KMnO_4$, $NaNO_3$, and H_2SO_4, successfully oxidising graphite to GrO within a few hours [115]. This is the most widely employed approach for synthesising GrO due to its relative rapidity and safety compared to alternative methods. Notably, it does not produce toxic gases, such as chlorine dioxide (ClO_2) or acidic fog, when utilising potassium permanganate ($KMnO_4$) and sodium nitrate ($NaNO_3$) as reagents, in contrast to potassium perchlorate ($KClO_3$) and nitric acid (HNO_3) [116].

Similarly, Marcano et al. [117] aimed to enhance the Hummers method by increasing the oxidation level of GrO while reducing toxic gas emissions (e.g. NO_2, N_2O_4). In this updated method, potassium permanganate replaced sodium nitrate, and the oxidation reaction used a mixture of H_2SO_4 and H_3PO_4 in a 9:1 ratio. Furthermore, modifications to the Hummers method have brought more environmentally benign approaches to GrO synthesis; for instance, removing $NaNO_3$ in the improved Hummers method has reduced production costs and environmental ramifications [8]. The oxidative treatment of graphite increases the interlayer spacing, expanding from 0.335 nm to 0.625 nm or more, attributable to the formation of intercalation compounds [118]. This alteration weakens van der Waals interactions among the layers, thereby facilitating the exfoliation into well-dispersed single, bi-, and few-layer GrO upon sonication in suitable solvents. GrO exhibits a hydrophilicity due to its oxygen functional groups, which enhances its dispersibility in a range of solvents, including water, ethylene glycol, N-methyl-2-pyrrolidone (NMP), and tetrahydrofuran (THF) [119, 120].

All the above-mentioned techniques use an acidic reaction environment and utilise potent oxidising agents, including $KClO_3$ or $KMnO_4$. The specific method chosen is critical, as it affects the yield of different functional groups (such as OH, COOH, ketones, and epoxides) and the overall degree of oxidation (the C/O ratio) [121–123]. However, some approaches examined amorphous and nonstoichiometric alternatives, e.g. He et al. [104] and Lerf et al. [108, 124] developed their model based on experimental findings from solid-state nuclear magnetic resonance and various X-ray analysis techniques, suggesting the potential for hydrogen bonding between GrO sheets.

Graphene research predominantly centres around GrO, an electrically insulating material characterised by reactive oxygen functionalities. In contrast to pristine graphene, which possesses a uniform structural configuration, GrO accommodates a diverse array of functional groups integrated into its framework. The structure of GrO comprises a combination of sp^2- and sp^3-hybridised carbon atoms, along with various oxygen-containing functional groups, including hydroxyl (C–OH) and epoxide (C–O–C) groups located on the basal plane and carbonyl ($C = O$) and carboxyl (COOH) groups at the edges of the sheets [124–128]. The presence of oxygen-containing functional groups enhances the hydrophilicity of GrO sheets, allowing them to swell and disperse readily in aqueous media. Following its exfoliation into individual sheets, the oxidation of graphite to synthesise GrO typically depends on the specific reaction conditions employed and yields a heterogeneous distribution of surface functionalities. Pristine graphene consists of sp^2-hybridised carbon atoms, while GrO encompasses a two-dimensional network that incorporates sp^2 and sp^3 hybridised regions. The oxidation process inherently introduces defects, impurities, and other structural alterations, which can significantly influence the electronic, optical, and adsorption properties of GrO. The functional groups introduce sp^3 defects within the nanosheets, disrupting the intrinsic conjugated π system and reducing strength and electrical conductivity. Due to sp^3-hybridised carbon atoms bonded to oxygen functional groups, GrO exhibits characteristics of an electronically hybrid material. This structural modification introduces an energy gap between the non-conductive σ-states of the sp^3-bonded carbon atoms and the conductive π-states associated with sp^2 carbon domains [100]. The ratio of sp^2 to sp^3 fractions can be adjusted by employing reduction chemistry, allowing for the controlled transformation of GrO from an insulator to a semiconductor or semi-metal [102].

2.4.2 *Reduced Graphene Oxide (rGrO)*

GrO can be transformed into graphene by recovering its conjugated structure and eliminating oxygen-containing groups. The resulting product, rGrO, is often called chemically modified or chemically transformed graphene [102]. GrO can be converted into rGrO through chemical, photochemical, thermal, microwave, and solvothermal processes, which diminish the oxygen functional groups present in GrO, thereby yielding rGrO with a carbon-to-oxygen (C/O) ratio ranging from 8:1 to 246:1 [97, 129]. Common reduction agents include hydrogen gas, vitamin C, hydrazine, bovine serum albumin, and strong alkaline solutions.

The physicochemical properties of rGrO increasingly approximate those of pristine graphene nanosheets due to the partial removal of oxygen-containing functional groups during the reduction process. This deoxygenation facilitates the re-establishment of sp^2 hybridised carbon networks and improves the material's structural order and electronic conductivity.

Ma et al. [130] analysed the FTIR spectra of rGrO and observed that not all functional groups present in GrO were entirely removed during the reduction process; thus, rGrO retains some functional groups such as -OH, -COOH, and -C = O. Other researchers have reported similar findings. Further analyses have indicated that although rGrO is a reduced form of GrO, it is expected to exhibit properties similar to those of graphene nanosheets. However, its characteristics can vary depending on the reduction conditions, resulting in properties that fall between those of Gro and graphene.

2.4.2.1 Thermal Reduction

The thermal reduction of GrO can be achieved through various methods, including conventional heating (thermal annealing) [131], microwave irradiation [132], and photo-reduction [133]. In thermal annealing, GrO is exposed to high temperatures in the presence of inert gases such as argon. The rapid increase in temperature leads to the breakdown of oxygen functional groups into carbon monoxide and carbon dioxide gases. This generates significant pressure between the GrO layers, separating the stacked GrO sheets [131].

Microwave irradiation (MWI) is another method for thermal reduction, as it employs a microwave oven to heat the GrO quickly and uniformly [132]. Furthermore, Zhang et al. introduced a photo-reduction technique that utilises femtosecond laser irradiation [133]. Guo et al. reported that [134] rGrO produced through photothermal reduction in air and nitrogen indicated improved thermal stability compared to GrO. However, it remains slightly less stable than graphite. Moreover, XPS studies confirmed a decrease in oxygen-containing groups in rGrO compared to GrO [134–136]. Thermal reduction poses challenges, including the necessity for high temperatures, which escalate operational costs, alongside the concomitant release of CO_2, a greenhouse gas. Conversely, hydrothermal reduction offers a practical way to reduce GrO at lower temperatures, requiring less energy [137].

2.4.2.2 Chemical Reduction

The methods that use chemical reducing agents are particularly advantageous for large-scale graphene production due to their accessibility and low cost compared to other reduction techniques [138]. The first instance of preparing rGrO was conducted by Stankovich et al., who utilised hydrazine as the reducing agent [139]. Both hydrazine and its derivatives, such as hydrazine hydrate and dimethylhydrazine, serve effectively as chemical reducing agents for GrO. In chemical reduction, hydrazine (N_2H_4) is a common reductant for lowering the oxygen content in GrO. However, its large-scale application is limited due to its expense and toxicity. Consequently, a more environmentally friendly approach has been sought to replace toxic N_2H_4 in GrO reduction [140, 141]. A variety of reducing agents have been investigated, including phosphorus. Borohydrides, aluminium hydride, hydrohalic acids, sulphur- and nitrogen-based reagents, oxygen-containing compounds, metal–acid and metal-alkaline mixtures, as well as amino acids [106].

Phosphorus can be effectively integrated into rGrO through a reaction involving GrO and white phosphorus or phosphine [142]. Furthermore, the response between GrO and phosphorus trihalogenide promotes the partial reduction of GrO. This also enables the simultaneous doping of phosphorus and halogen elements, thereby improving the material's intrinsic properties and functionality, as well as its potential for various applications within materials science [143].

Metal hydrides such as sodium borohydride, lithium aluminium hydride, and sodium hydride are potent reducing agents in organic chemistry. Notably, these metal hydrides exhibit strong reactivity with aqueous solutions [144], potentially complicating their application. Conversely, research by S. Some et al. (2013) introduced thiophene as a viable reducing agent for GrO, wherein thiophene donates electrons while simultaneously accepting oxygen [145]. This reduction process leads to an intermediate oxidised polymerised thiophene, which transforms by losing sulphur atoms, yielding a polyhydrocarbon. Upon heat treatment, this process results in the formation of rGrO.

Fernandez-Merino et al. examined several reducing agents, including sodium borohydride, pyro- gallol, and vitamin C, alongside hydrazine and found that vitamin C yielded the most effective suspension compared to the others [146]. Additionally, a study by Dang et al. (2012) assessed the impact of temperature on the reduction of GrO by hydrazine in a solvent mixture of N, N-dimethylformamide (DMF) and water, as well as the dispersibility of the resultant graphene in DMF. The results demonstrated that increased reduction temperatures resulted in a high carbon-to-oxygen (C/O) ratio and enhanced dispersibility, reducing the propensity for irreversible aggregation of graphene sheets. Specifically, the dispersibility of GrO decreased from 1.66 to 0.38 mg mL^{-1} as the reduction temperature was elevated from 25 $^{\circ}C$ to 80 $^{\circ}C$ [147].

Although chemical reduction techniques can attain a high C/O ratio with agents such as zinc/hydrochloric acid (33.5:1) [148], benzyl alcohol (30:1) [149], thionyl chloride (8.48:1) [150], caffeic acid (7.15:1) [151], sodium borohydride (6.9:1) [152], and baker's yeast (5.9:1) [153], challenges regarding safety, additional expenditures, environmental pollution, and elongated synthesis processes must be addressed for the large-scale feasibility of these methods. Researchers are increasingly exploring alternative reduction strategies, such as thermal and hydrothermal reductions [154].

2.4.2.3 Electrochemical Reduction

Electrochemical reduction is becoming more popular due to its speed, cost-effectiveness, environmental advantages, and less toxic reductants than chemical methods [155, 156]. The electrochemical reduction of GrO involves the removal of oxygen functional groups using conventional electrochemical cells filled with aqueous solutions (electrolytes) at room temperature [157, 158]. In this process, electrons are transferred between inert electrodes and the GrO sheets or films. This method eliminates the necessity for hazardous reducing agents such as hydrazine.

2.4.2.4 Green Synthesis

The green synthesis of graphene reduction is rapidly evolving due to its low cost and simplicity. This environmentally friendly method utilises non-hazardous and inexpensive chemicals [159]. Various biomolecules, microorganisms, and plant extracts serve as reducing agents and stabilisers in the synthesis of metal nanoparticles. Plants are abundant in diverse biomolecular compounds, including amino acids, proteins, alcohols, polysaccharides, enzymes, polyphenols, and vitamins, which can effectively act as reducing and stabilising agents for graphene synthesis [160, 161].

Ascorbic acid, also known as Vitamin C, emerged as a highly effective reducing agent for GrO, offering several advantages over hydrazine [146]. It is nontoxic and demonstrates excellent chemical stability in water, which helps prevent the aggregation of the resulting rGrO. Recent initiatives have focused on green reduction methods to produce rGrOs from GrO utilising ascorbic acid due to environmental and health considerations[146], glucose [162], fructose, sucrose [132], tea leaves [159], melatonin [161], and alcohol [149]. The production of GrO through the oxidative exfoliation of graphite, followed by reduction to rGrO, is associated with low costs and high yields. However, challenges such as van der Waals forces can lead to a need for greater surface area, enhanced electronic conductivity, low solubility, and irreversible restacking of sheets in rGrO [163].

Additionally, the quality of rGrO is substantially influenced by factors, including the nature of the reductant employed and the specific conditions under which the reduction occurs, such as time, temperature, pressure, and voltage. Moreover, the carbon-to-oxygen (C/O) ratio indicates the suitability of the reducing agent in GrO reduction. The larger the C/O ratio, the higher the degree of deoxygenation and the higher the quality of rGrO [106].

2.5 Summary

This chapter examines the various production processes of graphene and its derivatives. The production of graphene is broadly categorised into bottom-up and top-down approaches. The bottom-up method encompasses epitaxial growth and chemical vapour deposition (CVD), while the top-down approach includes exfoliation, ball milling, liquid phase exfoliation, chemical exfoliation, and electrochemical exfoliation. Two types of graphene derivatives are mainly oxidised graphene (GrO) and rGrO. The GrO can be synthesised using four primary methods: Brodie, Staudenmaier, Hofmann, and Hummers. rGrO can be produced through thermal reduction, chemical reduction, electrochemical reduction, and green synthesis. Each production method for graphene and its derivatives presents distinct advantages and disadvantages, which are also discussed in this chapter. The next chapter will address the biocompatibility of graphene.

References

1. M. Yi, Z. Shen, A review on mechanical exfoliation for the scalable production of graphene. J. Mater. Chem. A **3**(22), 11700–11715 (2015)
2. Y. Zhang, L. Zhang, C. Zhou, Review of chemical vapor deposition of graphene and related applications. Acc. Chem. Res. **46**(10), 2329–2339 (2013)
3. C. Mattevi, H. Kim, M. Chhowalla, A review of chemical vapour deposition of graphene on copper. J. Mater. Chem. **21**(10), 3324–3334 (2011)
4. H.K. Yu, K. Balasubramanian, K. Kim, J.-L. Lee, M. Maiti, C. Ropers, J. Krieg, K. Kern, A.M. Wodtke, Chemical vapor deposition of graphene on a "peeled-off" epitaxial cu (111) foil: A simple approach to improved properties. ACS Nano **8**(8), 8636–8643 (2014)
5. A. Guti´errez-Cruz, A.R. Ruiz-Hern´andez, J.F. Vega-Clemente, D.G. Luna-Gazc´on, J. Campos- Delgado, A review of top-down and bottom-up synthesis methods for the production of graphene, graphene oxide and reduced graphene oxide. J. Mater. Sci. **57**(31), 14543–14578 (2022)

6. J.N. Coleman, Liquid exfoliation of defect-free graphene. Acc. Chem. Res. **46**(1), 14–22 (2013)
7. A. Ciesielski, P. Samor'I, Graphene via sonication assisted liquid-phase exfoliation. Chem. Soc. Rev. **43**(1), 381–398 (2014)
8. Y.L. Zhong, Z. Tian, G.P. Simon, D. Li, Scalable production of graphene via wet chemistry: progress and challenges. Mater. Today **18**(2), 73–78 (2015)
9. M.S.A. Bhuyan, M.N. Uddin, M.M. Islam, F.A. Bipasha, S.S. Hossain, Synthesis of graphene. Int. Nano Lett. **6**(2), 65–83 (2016)
10. S.Y. Zhou, G.-H. Gweon, A. Fedorov, D.P.N First, W. De Heer, D.-H. Lee, F. Guinea, A. Castro Neto, A. Lanzara, Substrate-induced bandgap opening in epitaxial graphene. Nat. Mater. **6**(10), 770–775 (2007)
11. R. Van Noorden, Moving towards a graphene world. Nature **442**(7100), 228–230 (2006)
12. Z. Tehrani, G. Burwell, M.M.Azmi, A. Castaing, R. Rickman, J. Almarashi, P. Dunstan, A.M. Beigi, S. Doak, O. Guy, Generic epitaxial graphene biosensors for ultrasensitive detection of cancer risk biomarker. 2D Materials **1**(2), 025004 (2014)
13. J. Hass, W.D. Heer, E. Conrad,The growth and morphology of epitaxial multilayer graphene. J. Phys.: Condens. Matter **20**(32), 323202 (2008)
14. K.S. Novoselov, L. Colombo, P. Gellert, M. Schwab, K. Kim et al., A roadmap for graphene. nature **490**(7419), 192–200 (2012)
15. S. Hertel, D. Waldmann, J. Jobst, A. Albert, M. Albrecht, S. Reshanov, A. Sch¨oner, M. Krieger, H.B. Weber, Tailoring the graphene/silicon carbide interface for monolithic wafer-scale electronics. Nat. Commun. **3**(1), 957 (2012)
16. C. Berger, Z. Song, T. Li, X. Li, A.Y. Ogbazghi, R. Feng, Z. Dai, A.N. Marchenkov, E.H. Conrad, P.N. First et al., Ultrathin epitaxial graphite: 2d electron gas properties and a route toward graphene-based nanoelectronics. J. Phys. Chem. B **108**(52), 19912–19916 (2004)
17. E. Rollings, G.-H. Gweon, S. Zhou, B. Mun, J. McChesney, B. Hussain, A. Fedorov, P. First, W. De Heer, A. Lanzara, Synthesis and characterization of atomically thin graphite films on a silicon carbide substrate. J. Phys. Chem. Solids **67**(9–10), 2172–2177 (2006)
18. C. Berger, Z. Song, X. Li, X. Wu, N. Brown, C. Naud, D. Mayou, T. Li, J. Hass, A.N. Marchenkov et al., Electronic confinement and coherence in patterned epitaxial graphene. Science **312**(5777), 1191–1196 (2006)
19. X. Xu, Z. Zhang, J. Dong, D. Yi, J. Niu, M. Wu, L. Lin, R. Yin, M. Li, J. Zhou et al., Ultrafast epitaxial growth of metre-sized single-crystal graphene on industrial cu foil. Science bulletin **62**(15), 1074–1080 (2017)
20. M. Bahri, S.H. Gebre, M.A. Elaguech, F.T. Dajan, M.G. Sendeku, C. Tlili, D. Wang, Recent advances in chemical vapour deposition techniques for graphene-based nanoarchitectures: From synthesis to contemporary applications. Coord. Chem. Rev. **475**, 214910 (2023)
21. Z. Yan, Z. Peng, J.M. Tour, Chemical vapor deposition of graphene single crystals. Acc. Chem. Res. **47**(4), 1327–1337 (2014)
22. M. Li, D. Liu, D. Wei, X. Song, D. Wei, A.T.S. Wee, Controllable synthesis of graphene by plasma-enhanced chemical vapor deposition and its related applications. Adv. Sci. **3**(11), 1600003 (2016)
23. A. Mohsin, L. Liu, P. Liu, W. Deng, I.N. Ivanov, G. Li, O.E. Dyck, G. Duscher, J.R. Dunlap, K. Xiao et al., Synthesis of millimeter-size hexagon-shaped graphene single crystals on resolidified copper. ACS Nano **7**(10), 8924–8931 (2013)
24. H. Zhou, W.J. Yu, L. Liu, R. Cheng, Y. Chen, X. Huang, Y. Liu, Y. Wang, Y. Huang, X. Duan, Chemical vapour deposition growth of large single crystals of monolayer and bilayer graphene. Nat. Commun. **4**(1), 2096 (2013)
25. X. Wan, K. Chen, D. Liu, J. Chen, Q. Miao, J. Xu, High-quality large-area graphene from dehydrogenated polycyclic aromatic hydrocarbons. Chem. Mater. **24**(20), 3906–3915 (2012)
26. X. Chen, L. Zhang, S. Chen, Large area cvd growth of graphene. Synth. Met. **210**, 95–108 (2015)
27. C. Miao, *CVD Synthesis of Graphene and Graphene Bipolar Junction Transistor* (University of California, Los Angeles, Los Angeles, USA, 2011)

28. Q. Zheng, J.K. Kim, Q. Zheng, J.K. Kim, Synthesis, structure, and properties of graphene and graphene oxide. Graphene Transparent Conduct: Synth. Prop. Appl. 29–94 (2015)
29. Q. Yu, J. Lian, S. Siriponglert, H. Li, Y.P. Chen, S.S. Pei, Graphene segregated on ni surfaces and transferred to insulators. Appl. Phys. Lett. **93**(11) (2008)
30. P.R. Somani, S.P. Somani, M. Umeno, Planer nano-graphenes from camphor by cvd. Chem. Phys. Lett. **430**(1–3), 56–59 (2006)
31. X. Li, W. Cai, J. An, S. Kim, J. Nah, D. Yang, R. Piner, A. Velamakanni, I. Jung, E. Tutuc et al, Large-area synthesis of high-quality and uniform graphene films on copper foils. Science **324**(5932), 1312–1314 (2009)
32. S. Bae, H. Kim, Y. Lee, X. Xu, J.-S. Park, Y. Zheng, J. Balakrishnan, T. Lei, H. Ri Kim, Y.I. Song et al., Roll-to-roll production of 30-inch graphene films for transparent electrodes. Nat. Nanotechnol. **5**(8), 574–578 (2010)
33. T. Kobayashi, M. Bando, N. Kimura, K. Shimizu, K. Kadono, N. Umezu, K. Miyahara, S. Hayazaki, S. Nagai, Y. Mizuguchi et al., Production of a 100-m-long high-quality graphene transparent conductive film by roll-to-roll chemical vapor deposition and transfer process. Applied Physics Letters **102**(2) (2013)
34. T. Yamada, M. Ishihara, J. Kim, M. Hasegawa, S. Iijima, A roll-to-roll microwave plasma chemical vapor deposition process for the production of 294 mm width graphene films at low temperature. Carbon **50**(7), 2615–2619 (2012)
35. I. Vlassiouk, P. Fulvio, H. Meyer, N. Lavrik, S. Dai, P. Datskos, S. Smirnov, Large scale atmospheric pressure chemical vapor deposition of graphene. Carbon **54**, 58–67 (2013)
36. E.S. Polsen, D.Q. McNerny, B. Viswanath, S.W. Pattinson, A. John Hart, High-speed roll-to-roll manufacturing of graphene using a concentric tube cvd reactor. Sci. Rep. **5**(1), 1–12 (2015)
37. L. Lin, J. Li, H. Ren, A.L. Koh, N. Kang, H. Peng, H. Xu, Z. Liu, Surface engineering of copper foils for growing centimeter-sized single-crystalline graphene. ACS Nano **10**(2), 2922–2929 (2016)
38. T.H. Bointon, M.D. Barnes, S. Russo, M.F. Craciun,High quality monolayer graphene synthesized by resistive heating cold wall chemical vapor deposition. Adv. Mater. (Deerfield Beach, Fla.) **27**(28), 4200 (2015)
39. Z. Bo, Y. Yang, J. Chen, K. Yu, J. Yan, K. Cen, Plasma-enhanced chemical vapor deposition synthesis of vertically oriented graphene nanosheets. Nanoscale **5**(12), 5180–5204 (2013)
40. K. Jia, H. Ci, J. Zhang, Z. Sun, Z. Ma, Y. Zhu, S. Liu, J. Liu, L. Sun, X. Liu et al., Super- clean growth of graphene using a cold-wall chemical vapor deposition approach. Angew. Chem. **132**(39), 17367–17371 (2020)
41. W. Mu, Y. Fu, S. Sun, M. Edwards, L. Ye, K. Jeppson, J. Liu, Controllable and fast synthesis of bilayer graphene by chemical vapor deposition on copper foil using a cold wall reactor. Chem. Eng. J. **304**, 106–114 (2016)
42. S. Das, J. Drucker, Nucleation and growth of single layer graphene on electrodeposited cu by cold wall chemical vapor deposition. Nanotechnology **28**(10), 105601 (2017)
43. Y. Song, W. Zou, Q. Lu, L. Lin, Z. Liu, Graphene transfer: Paving the road for applications of chemical vapor deposition graphene. Small **17**(48), 2007600 (2021)
44. J. Hwang, M. Kim, D. Campbell, H.A. Alsalman, J.Y. Kwak, S. Shivaraman, A.R. Woll, A.K. Singh, R.G. Hennig, S. Gorantla et al., Van der waals epitaxial growth of graphene on sapphire by chemical vapor deposition without a metal catalyst. ACS Nano **7**(1), 385–395 (2013)
45. J. Chen, Y. Guo, Y. Wen, L. Huang, Y. Xue, D. Geng, B. Wu, B. Luo, G. Yu, Y. Liu, Two-stage metal-catalyst-free growth of high-quality polycrystalline graphene films on silicon nitride substrates. Adv. Mater. **25**(7), 992–997 (2013)
46. J. Chen, Y. Wen, Y. Guo, B. Wu, L. Huang, Y. Xue, D. Geng, D. Wang, G. Yu, Y. Liu, Oxygen-aided synthesis of polycrystalline graphene on silicon dioxide substrates. J. Am. Chem. Soc. **133**(44), 17548–17551 (2011)
47. K.S. Novoselov, A. Mishchenko, A. Carvalho, A. Castro Neto, 2d materials and van der waals heterostructures. Science **353**(6298), 9439 (2016)

48. S.S. Datta, D.R. Strachan, S.M. Khamis, A.C. Johnson, Crystallographic etching of few-layer graphene. Nano Lett. **8**(7), 1912–1915 (2008)
49. Y. Zhang, J.P. Small, W.V. Pontius, P. Kim, Fabrication and electric-field-dependent transport measurements of mesoscopic graphite devices. Appl. Phys. Lett. **86**(7) (2005)
50. K.S. Novoselov, A.K. Geim, S.V. Morozov, D.E. Jiang, Y. Zhang, S.V. Dubonos, I.V. Grigorieva, A.A. Firsov, Electric field effect in atomically thin carbon films. Science **306**(5696), 666–669 (2004)
51. A.K. Geim, Nobel lecture: Random walk to graphene. Rev. Mod. Phys. **83**(3), 851–862 (2011)
52. D.G. Papageorgiou, I.A. Kinloch, R.J. Young, Mechanical properties of graphene and graphene-based nanocomposites. Prog. Mater Sci. **90**, 75–127 (2017)
53. C. Soldano, A. Mahmood, E. Dujardin, Production, properties and potential of graphene. Carbon **48**(8), 2127–2150 (2010)
54. K.I. Bolotin, K. Sikes, Z. Jiang, M. Klima, G. Fudenberg, J. Hone, P. Kim, H.L. Stormer, Ultrahigh electron mobility in suspended graphene. Solid State Commun. **146**(9–10), 351–355 (2008)
55. J.-H. Chen, C. Jang, S. Xiao, M. Ishigami, M.S. Fuhrer, Intrinsic and extrinsic performance limits of graphene devices on sio2. Nat. Nanotechnol. **3**(4), 206–209 (2008)
56. M. Ishigami, J.-H. Chen, W.G. Cullen, M.S. Fuhrer, E.D. Williams, Atomic structure of graphene on sio2. Nano Lett. **7**(6), 1643–1648 (2007)
57. J. Moser, A. Barreiro, A. Bachtold, Current-induced cleaning of graphene. Appl. Phys. Lett. **91**(16) (2007)
58. M.V. Antisari, A. Montone, N. Jovic, E. Piscopiello, C. Alvani, L. Pilloni, Low energy pure shear milling: a method for the preparation of graphite nano-sheets. Scripta Mater. **55**(11), 1047–1050 (2006)
59. A. Milev, M. Wilson, G.K. Kannangara, N. Tran, X-ray diffraction line profile analysis of nanocrystalline graphite. Mater. Chem. Phys. **111**(2–3), 346–350 (2008)
60. C. Knieke, A. Berger, M. Voigt, R.N.K. Taylor, J. R¨ohrl, W. Peukert, Scalable production ofgraphene sheets by mechanical delamination. Carbon **48**(11), 3196–3204 (2010)
61. C. Damm, T.J. Nacken, W. Peukert, Quantitative evaluation of delamination of graphite by wet media milling. Carbon **81**, 284–294 (2015)
62. Y. Lv, L. Yu, C. Jiang, S. Chen, Z. Nie, Synthesis of graphene nanosheet powder with layer number control via a soluble salt-assisted route. RSC Adv. **4**(26), 13350–13354 (2014)
63. O.Y. Posudievsky, O.A. Khazieieva, V.V. Cherepanov, V.G. Koshechko, V.D. Pokhodenko, High yield of graphene by dispersant-free liquid exfoliation of mechanochemically delaminated graphite. J. Nanopart. Res. **15**, 1–9 (2013)
64. T. Lin, Y. Tang, Y. Wang, H. Bi, Z. Liu, F. Huang, X. Xie, M. Jiang, Scotch-tape- like exfoliation of graphite assisted with elemental sulfur and graphene–sulfur composites for high-performance lithium-sulfur batteries. Energy Environ. Sci. **6**(4), 1283–1290 (2013)
65. V. Le´on, A.M. Rodriguez, P. Prieto, M. Prato, E. Vazquez, Exfoliation of graphite with triazine derivatives under ball-milling conditions: preparation of few-layer graphene via selective noncovalent interactions. ACS nano **8**(1), 563–571 (2014)
66. K.R. Paton, E. Varrla, C. Backes, R.J. Smith, U. Khan, A. O'Neill, C. Boland, M. Lotya, O.M. Istrate, P. King et al., Scalable production of large quantities of defect-free few-layer graphene by shear exfoliation in liquids. Nat. Mater. **13**(6), 624–630 (2014)
67. Y. Hernandez, V. Nicolosi, M. Lotya, F.M. Blighe, Z. Sun, S. De, I. T. McGovern, B. Hol- land, M. Byrne, Y.K. Gun'Ko et al., High-yield production of graphene by liquid-phase exfoliation of graphite. Nature Nanotech **3**(9), 563–568 (2008)
68. A.M. Dimiev, G. Ceriotti, A. Metzger, N.D. Kim, J.M. Tour, Chemical mass production of graphene nanoplatelets in 100% yield. ACS Nano **10**(1), 274–279 (2016)
69. U. Khan, A. O'Neill, M. Lotya, S. De, J.N. Coleman, High-concentration solvent exfoliation of graphene. small **6**(7), 864–871 (2010)
70. A.B. Bourlinos, V. Georgakilas, R. Zboril, T.A. Steriotis, A.K. Stubos, Liquid-phase exfoliation of graphite towards solubilized graphenes. Small **5**(16), 1841–1845 (2009)

71. L. Zhu, X. Zhao, Y. Li, X. Yu, C. Li, Q. Zhang, High-quality production of graphene by liquid-phase exfoliation of expanded graphite. Mater. Chem. Phys. **137**(3), 984–990 (2013)
72. G.S. Bang, H.-M. So, M.J. Lee, C.W. Ahn, Preparation of graphene with few defects using expanded graphite and rose bengal. J. Mater. Chem. **22**(11), 4806–4810 (2012)
73. J.H. Lee, D.W. Shin, V.G. Makotchenko, A.S. Nazarov, V.E. Fedorov, Y.H. Kim, J.-Y. Choi, J.M. Kim, J.-B. Yoo, One-step exfoliation synthesis of easily soluble graphite and transparent conducting graphene sheets. Adv. Mater. **21**(43), 4383 (2009)
74. W. Gu, W. Zhang, X. Li, H. Zhu, J. Wei, Z. Li, Q. Shu, C. Wang, K. Wang, W. Shen et al., Graphene sheets from worm-like exfoliated graphite. J. Mater. Chem. **19**(21), 3367–3369 (2009)
75. N. Behabtu, J.R. Lomeda, M.J. Green, A.L. Higginbotham, A. Sinitskii, D.V. Kosynkin, D. Tsentalovich, A.N.G. Parra-Vasquez, J. Schmidt, E. Kesselman et al., Spontaneous high-concentration dispersions and liquid crystals of graphene. Nat. Nanotechnol. **5**(6), 406–411 (2010)
76. A. O'Neill, U. Khan, P.N. Nirmalraj, J. Boland, J.N. Coleman, Graphene dispersion and exfoliation in low boiling point solvents. J. Phys. Chem. C **115**(13), 5422–5428 (2011)
77. D. Parviz, S. Das, H.T. Ahmed, F. Irin, S. Bhattacharia, M.J. Green, Dispersions of non-covalently functionalized graphene with minimal stabilizer. ACS Nano **6**(10), 8857–8867 (2012)
78. M. Raji, N. Zari, R. Bouhfid et al., Chemical preparation and functionalization techniques of graphene and graphene oxide. In *Functionalized Graphene Nanocomposites and Their Derivatives*, pp. 1–20. Elsevier, Amsterdam, Netherlands (2019)
79. Y. Seekaew, O. Arayawut, K. Timsorn, C. Wongchoosuk, C.: Synthesis, characterization, and applications of graphene and derivatives. In *Carbon-based Nanofillers and Their Rubber Nanocomposites*, pp. 259–283. Elsevier, Amsterdam, Netherlands (2019)
80. L.M. Viculis, J.J. Mack, R.B. Kaner, A chemical route to carbon nanoscrolls. Science **299**(5611), 1361–1361 (2003)
81. K. Parvez, S. Yang, X. Feng, K. Mu¨llen, Exfoliation of graphene via wet chemical routes. Synthetic Metals **210**, 123–132 (2015)
82. R. Raccichini, A. Varzi, S. Passerini, B. Scrosati, The role of graphene for electrochemical energy storage. Nat. Mater. **14**(3), 271–279 (2015)
83. A. Abdelkader, A. Cooper, R.A. Dryfe, I. Kinloch, How to get between the sheets: a review of recent works on the electrochemical exfoliation of graphene materials from bulk graphite. Nanoscale **7**(16), 6944–6956 (2015)
84. M. Raji, N. Zari, A. El Kacem Qaiss, R. Bouhfid, *Functionalized Graphene Nanocomposites and Their Derivatives* (Elsevier, Amsterdam, Netherlands, 2019)
85. C. Low, F. Walsh, M. Chakrabarti, M. Hashim, M. Hussain, Electrochemical approaches to the production of graphene flakes and their potential applications. Carbon **54**, 1–21 (2013)
86. Y.F. Li, S.M. Chen, W.H. Lai, Y.J. Sheng, H.K. Tsao, Superhydrophilic graphite surfaces and water-dispersible graphite colloids by electrochemical exfoliation. J. Chem. Phys. **139**(6) (2013)
87. J. Lu, J.-X. Yang, J. Wang, A. Lim, S. Wang, K.P. Loh, One-pot synthesis of fluorescent carbon nanoribbons, nanoparticles, and graphene by the exfoliation of graphite in ionic liquids. ACS Nano **3**(8), 2367–2375 (2009)
88. K. Parvez, R. Li, S.R. Puniredd, Y. Hernandez, F. Hinkel, S. Wang, X. Feng, K. Mullen, Electrochemically exfoliated graphene as solution-processable, highly conductive electrodes for organic electronics. ACS Nano **7**(4), 3598–3606 (2013)
89. K. Parvez, Z.-S. Wu, R. Li, X. Liu, R. Graf, X. Feng, K. Mullen, Exfoliation of graphite into graphene in aqueous solutions of inorganic salts. J. Am. Chem. Soc. **136**(16), 6083–6091 (2014)
90. J. Wang, K.K. Manga, Q. Bao, K.P. Loh, High-yield synthesis of few-layer graphene flakes through electrochemical expansion of graphite in propylene carbonate electrolyte. J. Am. Chem. Soc. **133**(23), 8888–8891 (2011)

91. N. Liu, F. Luo, H. Wu, Y. Liu, C. Zhang, J. Chen, One-step ionic-liquid-assisted electrochemical synthesis of ionic-liquid-functionalized graphene sheets directly from graphite. Adv. Func. Mater. **18**(10), 1518–1525 (2008)
92. M. Mao, M. Wang, J. Hu, G. Lei, S. Chen, H. Liu, Simultaneous electrochemical synthesis of few-layer graphene flakes on both electrodes in protic ionic liquids. Chem. Commun. **49**(46), 5301–5303 (2013)
93. C.-Y. Su, A.-Y. Lu, Y. Xu, F.-R. Chen, A.N. Khlobystov, L.-J. Li, High-quality thin graphene films from fast electrochemical exfoliation. ACS Nano **5**(3), 2332–2339 (2011)
94. Y.L. Zhong, T.M. Swager, Enhanced electrochemical expansion of graphite for in situ electrochemical functionalization. J. Am. Chem. Soc. **134**(43), 17896–17899 (2012)
95. C.-J. Shih, A. Vijayaraghavan, R. Krishnan, R. Sharma, J.-H. Han, M.-H. Ham, Z. Jin, S. Lin, G.L. Paulus, N.F. Reuel et al., Bi-and trilayer graphene solutions. Nat. Nanotechnol. **6**(7), 439–445 (2011)
96. A. Ambrosi, M. Pumera, Electrochemically exfoliated graphene and graphene oxide for energy storage and electrochemistry applications. Chemistry–A Eur. J. **22**(1), 153–159 (2016)
97. M. Tahriri, M. Del Monico, A. Moghanian, M.T. Yaraki, R. Torres, A. Yadegari, L. Tayebi, Graphene and its derivatives: Opportunities and challenges in dentistry. Mater. Sci. Eng., C **102**, 171–185 (2019)
98. V. Rosa, H. Xie, N. Dubey, T.T. Madanagopal, S.S. Rajan, J.L.P. Morin, I. Islam, A.H.C. Neto, Graphene oxide-based substrate: physical and surface characterization, cytocompati- bility and differentiation potential of dental pulp stem cells. Dent. Mater. **32**(8), 1019–1025 (2016)
99. N. Dubey, R. Bentini, I. Islam, T. Cao, A.H. Castro Neto, V. Rosa, Graphene: a versatile carbon-based material for bone tissue engineering. Stem Cells Int. **2015**(1), 804213 (2015)
100. O.C. Compton, S.T. Nguyen, Graphene oxide, highly reduced graphene oxide, and graphene: versatile building blocks for carbon-based materials. Small **6**(6), 711–723 (2010)
101. D.R. Dreyer, S. Park, C.W. Bielawski, R.S. Ruoff, The chemistry of graphene oxide. Chem. Soc. Rev. **39**(1), 228–240 (2010)
102. G. Eda, M. Chhowalla, Chemically derived graphene oxide: towards large-area thin-film electronics and optoelectronics. Adv. Mater. **22**(22), 2392–2415 (2010)
103. K.P. Loh, Q. Bao, G. Eda, M. Chhowalla, Graphene oxide as a chemically tunable platform for optical applications. Nat. Chem. **2**(12), 1015–1024 (2010)
104. H. He, J. Klinowski, M. Forster, A. Lerf, A new structural model for graphite oxide. Chem. Phys. Lett. **287**(1–2), 53–56 (1998)
105. T. Nakajima, Y. Matsuo, Formation process and structure of graphite oxide. Carbon **32**(3), 469–475 (1994)
106. C.K. Chua, M. Pumera, Chemical reduction of graphene oxide: a synthetic chemistry viewpoint. Chem. Soc. Rev. **43**(1), 291–312 (2014)
107. T. Szab´o, O. Berkesi, P. Forg´o, K. Josepovits, Y. Sanakis, D. Petridis, I. D´ek´any, Evolution of surface functional groups in a series of progressively oxidized graphite oxides. Chem. Mater. **18**(11), 2740–2749 (2006)
108. A. Lerf, H. He, T. Riedl, M. Forster, J. Klinowski, 13c and 1h mas nmr studies of graphite oxide and its chemically modified derivatives. Solid State Ionics **101**, 857–862 (1997)
109. H. Yu, B. Zhang, C. Bulin, R. Li, R. Xing, High-efficient synthesis of graphene oxide based on improved hummers method. Sci. Rep. **6**(1), 36143 (2016)
110. T. Somanathan, K. Prasad, K. Ostrikov, A. Saravanan, V. Mohana Krishna, Graphene oxide synthesis from agro waste. Nanomaterials **5**(2), 826–834 (2015)
111. B.C. Brodie, Xiii on the atomic weight of graphite. Philosophical transactions of the Royal Society of London (149), 249–259 (1859)
112. A. Moosa, M. Abed, Graphene preparation and graphite exfoliation. Turk. J. Chem. **45**(3), 493–519 (2021)
113. L. Staudenmaier, Verfahren zur darstellung der graphits¨aure. Berichte der deutschen chemischen Gesellschaft **31**(2), 1481–1487 (1898)
114. U. Hofmann, E. K¨onig, Untersuchungen u¨ber graphitoxyd. Zeitschrift fu¨r anorganische und allgemeine Chemie **234**(4), 311–336 (1937)

115. W.S. Hummers Jr., R.E. Offeman, Preparation of graphitic oxide. J. Am. Chem. Soc. **80**(6), 1339–1339 (1958)
116. J. Chen, B. Yao, C. Li, G. Shi, An improved hummers method for eco-friendly synthesis of graphene oxide. Carbon **64**, 225–229 (2013)
117. D.C. Marcano, D.V. Kosynkin, J.M. Berlin, A. Sinitskii, Z. Sun, A. Slesarev, L.B. Alemany, W. Lu, J.M. Tour, Improved synthesis of graphene oxide. ACS Nano **4**(8), 4806–4814 (2010)
118. G. Yoon, D.-H. Seo, K. Ku, J. Kim, S. Jeon, K. Kang, Factors affecting the exfoliation of graphite intercalation compounds for graphene synthesis. Chem. Mater. **27**(6), 2067–2073 (2015)
119. D. Konios, M.M. Stylianakis, E. Stratakis, E. Kymakis, Dispersion behaviour of graphene oxide and reduced graphene oxide. J. Colloid Interface Sci. **430**, 108–112 (2014)
120. M.S. Khan, A. Shakoor, G.T. Khan, S. Sultana, A. Zia, A study of stable graphene oxide dispersions in various solvents. J. Chem. Soc. Pak. **37**(1) (2015)
121. D. Lee, V.L. DeLos Santos, J. Seo, L.L. Felix, D.A. Bustamante, J. Cole, C. Barnes, The structure of graphite oxide: investigation of its surface chemical groups. J. Phys. Chem. B **114**(17), 5723–5728 (2010)
122. W. Gao, L.B. Alemany, L. Ci, P.M. Ajayan, New insights into the structure and reduction of graphite oxide. Nat. Chem. **1**(5), 403–408 (2009)
123. O. Jankovsky', P. Marvan, M. Nov´aˇcek, J. Luxa, V. Mazanek, K. Kl´ımov´a, D. Sedmidubsky', Z. Sofer,Synthesis procedure and type of graphite oxide strongly influence resulting graphene properties. Appl. Mater. Today 4, 45–53 (2016)
124. A. Lerf, H. He, M. Forster, J. Klinowski, Structure of graphite oxide revisited. J. Phys. Chem. B **102**(23), 4477–4482 (1998)
125. Z. Liu, X. Zhang, X. Yan, Y. Chen, J. Tian, Nonlinear optical properties of graphene-based materials. Chin. Sci. Bull. **57**, 2971–2982 (2012)
126. P. Bhawal, S. Ganguly, T. Chaki, N. Das, Synthesis and characterization of graphene oxide filled ethylene methyl acrylate hybrid nanocomposites. RSC Adv. **6**(25), 20781–20790 (2016)
127. D.R. Dreyer, R.S. Ruoff, C.W. Bielawski, From conception to realization: an historial account of graphene and some perspectives for its future. Angew. Chem. Int. Ed. **49**(49), 9336–9344 (2010)
128. K.P. Loh, Q. Bao, P.K. Ang, J. Yang, The chemistry of graphene. J. Mater. Chem. **20**(12), 2277–2289 (2010)
129. Y. Shang, D. Zhang, Y. Liu, C. Guo, Preliminary comparison of different reduction methods of graphene oxide. Bull. Mater. Sci. **38**, 7–12 (2015)
130. D. Ma, X. Li,Y. Guo, Y. Zeng, Study on ir properties of reduced graphene oxide. In: IOP Conference Series: Earth and Environmental Science, vol. 108, p. 022019. IOP Publishing (2018)
131. M.J. McAllister, J.L. Li, D.H. Adamson, H.C. Schniepp, A.A. Abdala, J. Liu, M. Herrera-Alonso, D.L. Milius, R. Car R.K. Prud'homme et al, Single sheet functionalized graphene by oxidation and thermal expansion of graphite. Chem. Mater. **19**(18), 4396–4404 (2007)
132. Y. Zhu, S. Murali, M.D. Stoller, A. Velamakanni, R.D. Piner, R.S. Ruoff, Microwave assisted exfoliation and reduction of graphite oxide for ultracapacitors. Carbon **48**(7), 2118–2122 (2010)
133. Y. Zhang, L. Guo, S. Wei, Y. He, H. Xia, Q. Chen, H.-B. Sun, F.-S. Xiao, Direct imprinting of microcircuits on graphene oxides film by femtosecond laser reduction. Nano Today **5**(1), 15–20 (2010)
134. H. Guo, M. Peng, Z. Zhu, L. Sun, Preparation of reduced graphene oxide by infrared irradiation induced photothermal reduction. Nanoscale **5**(19), 9040–9048 (2013)
135. F. Priante, M. Salim, L. Ottaviano, F. Perrozzi, Xps study of graphene oxide reduction induced by (100) and (111)-oriented si substrates. Nanotechnology **29**(7), 075704 (2018)
136. C. Xu, X. Shi, A. Ji, L. Shi, C. Zhou, Y. Cui, Fabrication and characteristics of reduced graphene oxide produced with different green reductants. PLoS ONE **10**(12), 0144842 (2015)
137. W. Chen, L. Yan, Preparation of graphene by a low-temperature thermal reduction at atmosphere pressure. Nanoscale **2**(4), 559–563 (2010)

138. S. Pei, H.-M. Cheng, The reduction of graphene oxide. Carbon **50**(9), 3210–3228 (2012)
139. S. Stankovich, D.A. Dikin, R.D. Piner, K.A. Kohlhaas, A. Kleinhammes, Y. Jia, Y. Wu, S.T. Nguyen, R.S. Ruoff, Synthesis of graphene-based nanosheets via chemical reduction of exfoliated graphite oxide. carbon **45**(7), 1558–1565 (2007)
140. o. Akhavan, R. Azimirad, H. Gholizadeh, F. Ghorbani, Hydrogen-rich water for green reduction of graphene oxide suspensions. Int. J. Hydrog. Energy **40**(16), 5553–5560 (2015)
141. X. Zhang, K. Li, H. Li, J. Lu, Q. Fu, Y. Chu, Graphene nanosheets synthesis via chemical reduction of graphene oxide using sodium acetate trihydrate solution. Synth. Met. **193**, 132–138 (2014)
142. H.L. Poh, Z. Sofer, M. Nov´aˇcek, M. Pumera, Concurrent phosphorus doping and reduction of graphene oxide. Chemistry–A Eur. J. **20**(15), 4284–4291 (2014)
143. L. Wang, Z. Sofer, R. Zboril, K. Cepe, M. Pumera, Phosphorus and halogen co-doped graphene materials and their electrochemistry. Chemistry–A Eur. J. **22**(43), 15444–15450 (2016)
144. H.J. Shin, K.K. Kim, A. Benayad, S.M. Yoon, H.K. Park, I.S. Jung, M.H. Jin, H.K. Jeong, J.M. Kim, JY. Choi et al, Efficient reduction of graphite oxide by sodium borohydride and its effect on electrical conductance. Adv. Funct. Mater. **19**(12), 1987–1992 (2009)
145. S. Some, Y. Kim, Y. Yoon, H. Yoo, S. Lee, Y. Park, H. Lee, High-quality reduced graphene oxide by a dual-function chemical reduction and healing process. Sci. Rep. **3**(1), 1929 (2013)
146. M.J. Fern´andez-Merino, L. Guardia, J. Paredes, S. Villar-Rodil, P. Solis-Fernandez, A. Martinez-Alonso, J. Tasc´on, Vitamin c is an ideal substitute for hydrazine in the reduction of graphene oxide suspensions. J. Phys. Chem. C **114**(14), 6426–6432 (2010)
147. T.T. Dang, V.H. Pham, S.H. Hur, E.J. Kim, B.-S. Kong, J.S. Chung, Superior dispersion of highly reduced graphene oxide in n, n-dimethylformamide. J. Colloid Interface Sci. **376**(1), 91–96 (2012)
148. X. Mei, J. Ouyang, Ultrasonication-assisted ultrafast reduction of graphene oxide by zinc powder at room temperature. Carbon **49**(15), 5389–5397 (2011)
149. D.R. Dreyer, S. Murali, Y. Zhu, R.S. Ruoff, C.W. Bielawski, Reduction of graphite oxide using alcohols. J. Mater. Chem. **21**(10), 3443–3447 (2011)
150. W. Chen, L. Yan, P. Bangal, Chemical reduction of graphene oxide to graphene by sulfur containing compounds. J. Phys. Chem. C **114**(47), 19885–19890 (2010)
151. Z. Bo, X. Shuai, S. Mao, H. Yang, J. Qian, J. Chen, J. Yan, K. Cen, Green preparation of reduced graphene oxide for sensing and energy storage applications. Sci. Rep. **4**(1), 4684 (2014)
152. L.G. Guex, B. Sacchi, K.F. Peuvot, R.L. Andersson, A.M. Pourrahimi, V. Str¨om, S. Farris, R.T. Olsson, Experimental review: chemical reduction of graphene oxide (go) to reduced graphene oxide (rgo) by aqueous chemistry. Nanoscale **9**(27), 9562–9571 (2017)
153. P. Khanra, T. Kuila, N.H. Kim, S.H. Bae, D.-S. Yu, J.H. Lee, Simultaneous bio- functionalization and reduction of graphene oxide by baker's yeast. Chem. Eng. J. **183**, 526–533 (2012)
154. X. Zheng, Y. Peng, Y. Yang, J. Chen, H. Tian, X. Cui, W. Zheng, Hydrothermal reduction of graphene oxide; effect on surface-enhanced raman scattering. J. Raman Spectrosc. **48**(1), 97–103 (2017)
155. D. Voiry, J. Yang, J. Kupferberg, R. Fullon, C. Lee, H.Y. Jeong, H.S. Shin, M. Chhowalla, High-quality graphene via microwave reduction of solution-exfoliated graphene oxide. Science **353**(6306), 1413–1416 (2016)
156. V. Abdelsayed, S. Moussa, H.M. Hassan, H.S. Aluri, M.M. Collinson, M.S. El-Shall, Photothermal deoxygenation of graphite oxide with laser excitation in solution and graphene-aided increase in water temperature. J. Phys. Chem. Lett. **1**(19), 2804–2809 (2010)
157. M. Zhou, Y. Wang, Y. Zhai, J. Zhai, W. Ren, F. Wang, S. Dong, Controlled synthesis of large-area and patterned electrochemically reduced graphene oxide films. Chemistry–A Eur. J. **15**(25), 6116–6120 (2009)
158. Z. Wang, X. Zhou, J. Zhang, F. Boey, H. Zhang, Direct electrochemical reduction of single-layer graphene oxide and subsequent functionalization with glucose oxidase. J. Phys. Chem. C **113**(32), 14071–14075 (2009)

159. A.A. Moosa, J.N. Jaafar, Green reduction of graphene oxide using tea leaves extract with applications to lead ions removal from water. Nanosci. Nanotechnol. **7**(2), 38–47 (2017)
160. S. Iravani, Green synthesis of metal nanoparticles using plants. Green Chem. **13**(10), 2638–2650 (2011)
161. A. Esfandiar, O. Akhavan, A. Irajizad, Melatonin as a powerful bio-antioxidant for reduction of graphene oxide. J. Mater. Chem. **21**(29), 10907–10914 (2011)
162. O. Akhavan, E. Ghaderi, S. Aghayee, Y. Fereydooni, A. Talebi, The use of a glucose-reduced graphene oxide suspension for photothermal cancer therapy. J. Mater. Chem. **22**(27), 13773–13781 (2012)
163. Y.S. Yun, G. Yoon, M. Park, S.Y. Cho, H.-D. Lim, H. Kim, Y.W. Park, B.H. Kim, K. Kang, H.-J. Jin, Restoration of thermally reduced graphene oxide by atomic-level selenium doping. NPG Asia Materials **8**(12), 338–338 (2016)

Chapter 3
Biocompatibility of Graphene-Based Materials

Abstract This chapter provides a comprehensive evaluation of the biocompatibility of graphene-based materials (GBMs), with a particular focus on their interactions with biological systems and their potential in biomedical applications. Emphasis is placed on understanding how physicochemical properties—such as surface chemistry, functionalisation, morphology, and dispersibility—influence cellular responses, biodistribution, and overall toxicological profiles. Key advances in tissue engineering are explored, highlighting how GBMs can support stem cell differentiation, promote tissue regeneration, and serve as scaffolds for complex biological architectures. The text draws on a wide range of both in vitro and in vivo studies to illustrate the nuanced and sometimes contradictory findings regarding cytotoxicity, immunogenicity, and long-term integration in living systems. Ongoing challenges, including standardisation of testing protocols, dose-dependent effects, and long-term safety, are critically examined. The discussion aims to clarify the current limitations while identifying promising strategies for enhancing the biocompatibility and functional performance of GBMs in clinical settings. By situating GBMs at the intersection of nanotechnology and regenerative medicine, this chapter contributes to a growing body of knowledge essential for the safe and effective deployment of advanced materials in biomedical engineering.

Keywords Graphene-based materials · Biocompatibility · Toxicology · Cellular interaction · Cytotoxicity · Hemocompatibility · Biodegradability

3.1 Introduction

Biocompatibility refers to the ability of materials to interact with cells, tissues, or the body without causing harm [1]. The importance of evaluating the biocompatibility of biomedical sensors is critical [2]. The biocompatibility of graphene is the primary consideration when using it in biomedical applications [3]. The prominence of graphene and graphene-based nanomaterials in biomedical applications necessitates a thorough evaluation to determine their biocompatibility [4]. It must

S. Debnath et al., *Graphene in Wearable Sensors for Health Monitoring*,
SpringerBriefs in Applied Sciences and Technology,
https://doi.org/10.1007/978-981-96-8850-0_3

demonstrate biocompatibility to prevent adverse reactions within living tissues [5]. Assessing the toxicological aspects of graphene is important before confirming the practical function in vivo [6–8]. It should avoid harmful effects on tissue engineering and facilitate effective tissue engineering techniques that can lead to successful stem cell differentiation along desired tissue lineages, indicating that the material possesses the necessary biocompatibility [9–13].

This biocompatibility must be demonstrable in both in vitro and in vivo environments. Graphene-based materials (GBMs) in tissue engineering have shown notable progress, demonstrating the biocompatibility of graphene-based materials, as supported by research studies [3]. However, it is important to note that different physicochemical properties, such as size and distribution, surface area, layer number, impurities, and many more affect its toxicity and interaction with biological systems and cells [1, 2, 14–17]. For example, mitochondrial function assays, such as the MTT (3-(4,5-dimethylthiazolyl-2)-2,5-diphenyltetrazolium bromide) assay, reveal that various shapes of graphene sheets and single wall carbon nanotubes (SWCNTs) exhibit divergent levels of toxicity—specifically, few-layer graphene (FLG) sheets demonstrate higher toxicity than SWCNTs at lower concentrations [18, 19]. Interestingly, higher concentrations of graphene have been found to significantly reduce its toxic effects [6]. Moreover, it has been established that surface functionalisation significantly improves the biocompatibility of graphene-based materials (GBMs) [20]. Conversely, inadequate regulation may result in harm to healthy cells, as demonstrated by the inflammatory responses triggered by micro-sized GrO in comparison to the nano-sized counterpart following uptake by macrophages [2]. Several research studies have been carried out on the biocompatibility of graphene [21–24]. Graphene-family materials have shown excellent biocompatibility with antibacterial [2, 25–28], anti-hemolytic [14, 29–31], and nontoxic [2, 32] effects. It is reported that GBMs exhibit antibacterial properties through physical destruction, lipid extraction, and photothermic impact, causing damage to bacterial cells, leading to their inactivation [33]. In vitro investigations have shown that GBMs exhibit higher biocompatibility when present in lower concentrations and are well-dispersed [2].

3.2 Graphene-Stem Cells Interactions

Stem cells represent a distinct category of cells known for their remarkable ability to develop into specialised types necessary to self-repair various tissues and organs within the human body [34]. GBMs are considered some of the most promising substances for applications in tissue engineering and regenerative medicine due to their high specific surface area, exceptional chemical stability, biocompatibility, and versatility in functionalisation. These properties are particularly advantageous for promoting stem cell growth, proliferation, and differentiation. Numerous studies suggest that graphene nanomaterials, owing to their surface chemistry and remarkable mechanical properties, can promote the growth, proliferation, and differentiation of mesenchymal stem cells (MSCs), neural stem cells (NSCs), and induced pluripotent stem cells (iPSCs) [35, 36] into diverse tissue types. Additionally, the unique behaviours of stem cells can be influenced by modifying the various surface properties of these graphene nanomaterials.

3.3 Graphene-MSCs Interactions

Studies have confirmed that osteoblastic cells and mesenchymal stem cell (MSC) cultures remain viable when grown on graphene substrates [37]. Furthermore, graphene displayed high biocompatibility, with cells retaining non-cytotoxic adhesion, migratory capacity, and phenotypic functionality. Research suggests that MSCs exposed to graphene exhibit distinct integrin heterodimers and demonstrate a unique arrangement of bone-specific extracellular matrix (ECM) proteins, such as fibronectin, collagen I, and vitronectin [38]. Furthermore, MSCs can differentiate into bone-forming cells under specific physical or mechanical cues, even without chemical inducers. This indicates that the physicochemical properties of graphene nanoparticles are safe and effective in promoting the production of bone-specific extracellular matrix (ECM). Studies on MSCs have demonstrated that graphene nanoplatelets, specifically UGZ–1004, can induce alterations in gene expression associated with self-renewal and cellular proliferation [39]. MSCs treated with UGZ–1004 retained their characteristic surface markers. Notably, UGZ–1004 considerably increased the expression of genes crucial for regulating the cell cycle and DNA repair, including Cyclin-Dependent Kinase 1 (CDK1), Cyclin-Dependent Kinase 2(CDK2), and Mouse Double Minute 2 Homolog (MDM2). The gene expression pattern suggested that UGZ–1004 could increase the self-renewal capabilities of MSCs and ensure robust cellular function and longevity. It has been reported that UGZ–1004 downregulates genes associated with tumour development, potentially reducing tumorigenic risks. These findings highlight the ability of UGZ–1004 to strengthen the proliferation and self-renewal capacity of mesenchymal stem cells (MSCs). This enhancement is crucial for advancing regenerative medicine applications.

3.4 Graphene-HMSCs Interactions

The transplantation of human mesenchymal stem cells (hMSCs) and their directed differentiation into bone, fat, and cartilage have been demonstrated through various biomaterials [40, 41]. Over the years, graphene has emerged as one of the most promising biocompatible scaffolds for promoting the growth and differentiation of hMSCs [42, 43]. It was reported that graphene facilitated the proliferation of hMSCs and accelerated their differentiation into osteoblasts [42]. This acceleration in osteogenesis occurred without the need for additional growth factors, and the differentiation rate was comparable to that achieved with standard growth factors. While it is clear that graphene promotes the differentiation of hMSCs toward an osteogenic lineage, the precise mechanisms underlying this enhanced differentiation were unclear.

However, motivated by the abovementioned observation, Lee et al. (2011) investigated to elucidate the correlation between the observed phenomena and the chemical properties of graphene [43]. Human bone marrow-derived mesenchymal stem cells (MSCs) were initially cultured on polydimethylsiloxane (PDMS), graphene (Gr), and GrO.

The study demonstrated that graphene's capacity to localise growth factors and differentiation agents, mediated by non-covalent $\pi\pi$ interactions, enhances its utility as a bioactive substrate [43]. Specific serum proteins and osteogenic inducers were observed to stack effectively on graphene, which enhanced stem cell adhesion, proliferation, and bone differentiation. The π π binding capability of graphene inhibited the adipogenic differentiation of hMSCs and led to denaturation [43].

Conversely, stem cell differentiation towards the adipogenic lineage remained unaffected in the presence of hydrophilic GrO. This phenomenon can be attributed to the disruption of extensive sp^2 conjugation and π π binding capabilities by the oxygen functional groups present on GrO, preventing insulin's denaturation upon adsorption. Consequently, the strong affinity of GrO for insulin promoted adipogenesis in hMSCs. Thus, the extent of π π stacking and the hydrogen bonding and electrostatic interactions facilitated by graphene and GrO played a critical role in directing hMSC differentiation towards various tissue lineages [43, 44].

3.5 Graphene-NSCs Interactions

The distinct surface characteristics and adaptable chemistry of graphene have demonstrated their ability to influence the differentiation of stem cells. When human neural stem cells (NSCs) were cultured on graphene, they differentiated more rapidly than those cultured on glass [45].

The differentiated NSCs exhibited elongated morphologies and distinct neurite outgrowths, with the formation of neural networks observed on the graphene substrate. This research emphasised how the graphene microenvironment enhances the differentiation of NSCs into neuronal lineage cells, particularly neurons. Guo et al. (2016) investigated the mechanistic basis for graphene's enhancement of neural stem cell (NSC) proliferation and differentiation. They reported that this accelerative effect on NSC development stems from graphene's considerable modulation of the cells' active and passive membrane bioelectric characteristics. However, during maturation, the NSCs cultured on graphene displayed enhanced action potential firing and overall synaptic activity [46].

3.6 Graphene-IPSCs Interactions

Graphene has been identified as an effective biocompatible substrate for the growth and differentiation of induced pluripotent stem cells (iPSCs) [47]. Specifically, hydrophilic GrO promotes better attachment and faster growth of iPSCs due to the oxide groups on its surface. Regarding differentiation, graphene promotes the spontaneous development of iPSCs into mesodermal and ectodermal lineages. Conversely, graphene inhibits the differentiation of iPSCs into the endoderm lineage, while GrO enhances this process, likely due to the hydroxyl groups present on its surface [44]. These cells can differentiate into nearly all cell types across the three germ layers, similar to embryonic stem cells [48]. Research conducted by Chen et al. (2012) found that GrO can considerably influence the differentiation potential of iPSCs, particularly towards endodermal lineages. The study investigated the impact of graphene and GrO as two-dimensional substrates on iPSC differentiation. The results indicated that substrates coated with graphene inhibited iPSC differentiation, whereas those with GrO significantly promoted it, especially towards the endodermal lineage. By day 9, there was a notable reduction in the expression levels of pluripotency markers Nanog and Oct4 compared to the control and graphene groups. Conversely, a marked upregulation of endodermal markers Gata4 and Ihh was observed. The authors proposed that surface functional groups on GrO may interact with membrane receptors of induced pluripotent stem cells (iPSCs), potentially triggering spontaneous differentiation via receptor-mediated signaling [47].

3.7 Antibacterial Activity of Graphene

The antibacterial properties of graphene and its nanomaterials are influenced by several factors, including exposure time, concentration, physical and chemical characteristics, and the tested bacteria types [49, 50].

Graphene nanocomposites demonstrate effective antibacterial action against both Gram-positive and Gram-negative bacteria. This effectiveness stems from graphene's ability to physically damage microorganisms by penetrating and disrupting their cell membranes, wrapping around cells, inducing mechanical stress, and removing phospholipids from lipid membranes. Additionally, graphene induces oxidative stress by generating reactive oxygen species (ROS) and facilitating charge transfer effects [4]. Graphene nanosheets can be integrated into bacterial membranes, leading to interactions where Van der Waals forces and graphene's hydrophobic nature extract phospholipids from bacterial lipid layers, causing irreparable damage [51–53].

Aggregated graphene sheets suspended in solution can isolate bacteria from nutrient-rich environments [52], as GrO nanosheets are easily functionalised and well-dispersed in water. However, the antibacterial activity of GrO nanosheets is largely size-dependent, with larger GrO sheets exhibiting more pronounced antibacterial effects against E. coli [54, 55]. Both graphene and its composites can effectively target bacterial biofilms and individual bacteria [56]. As mentioned previously, the antibacterial effect of graphene on these bacteria is dose-dependent; high concentrations of GrO can inhibit biofilm formation in both Gram-positive and Gram-negative bacteria, while lower concentrations may promote it, leading to counterproductive outcomes [57].

It was reported that GrO concentrations below 50 μg.mL^{-1} in nutrient media did not exert antimicrobial effects and may promote bacterial growth by facilitating biofilm formation. Conversely, combining GrO with polyoxyalkyleneamine at the same concentration can demonstrate antibacterial properties when bacteria are cultured in a phosphate-buffered saline solution [6, 58].

3.8 Graphene-Skin Interaction

Researchers have investigated the applications of graphene-based wearable sensors for skin health monitoring, which is a primary defence against environmental exposure [2, 59]. One of the key benefits of incorporating biocompatibility into wearable sensor design is that it enables prolonged use without causing discomfort or skin irritation. This characteristic enhances user experience and broadens the range of potential applications of wearable sensors [60]. However, limited knowledge exists regarding the toxicological data and cutaneous toxicity of GBMs on the skin [61]. If there is cutaneous exposure, it is likely to cause skin irritation and an allergic reaction. Additionally, GBMs can interact with skin proteins [62]. From the simulation, it was found that the presence of oxygen groups in GrO leads to steric effects and electrostatic repulsion, resulting in a lack of tighter contact between the skin keratin and the graphene basal plane. In contrast, the presence of electrostatic attraction and hydrogen bonding enhances the interaction with positively charged amino acid residues, such as lysine and arginine. It was reported that the biocompatibility of GrO is better than that of pristine graphene. The in vitro studies examined the cytotoxicity of GBMs toward skin keratinocytes and fibroblasts [63]. Research by Liao et al. [30] demonstrated that aggregated graphene sheets showed higher cytotoxicity towards human skin fibroblasts compared to reversibly aggregated GrO, which was attributed to their greater tendency to form aggregates. Another study [61] suggested that human primary keratinocytes could only be penetrated by GrO when exposed to high concentrations and for an extended period, such as those obtained by ball-milling treatment. This could harm mitochondrial activity associated with plasma membrane damage, indicating low cytotoxicity to both human keratinocytes and fibroblasts. Additionally, studies have shown that graphene sheets pose a higher toxicity risk to human skin fibroblast cells than GrO. This is due to the faster sedimentation and more compact aggregation formation of graphene sheets, which can hinder the absorption of nutrients and the growth of human skin fibroblasts. To further investigate the interaction between macrophages and few-layer (1-10 layers) graphene nanoplatelets (GPs), THP-1, a differentiated monocytic cell line, was treated with GPs. The results showed that THP-1 cells could only partially phagocytose GPs, indicating a phenomenon called frustrated phagocytosis [64]. Additionally, a dose-dependent effect of GPs on the membrane integrity and ROS level of THP-1 cells was observed.

A study by Zhang et al. [65] investigated the biocompatibility of a wearable hydrogel-based sensor composed of a polyvinyl alcohol-pristine graphene oxide-polydopamine (PVA-prGrO-PDA) composite. The hydrogel's cell toxicity was evaluated using the Cell Counting Kit-8 (CCK-8) assay in an in vitro cytotoxicity test, with NIH3T3 fibroblasts cultured for 24, 36, and 48 hours during the test. The results showed that the fibroblasts cultured for 48 hours in soaking solution substrates of the PVA-prGrO-PDA hydrogel displayed an irregular triangle morphology, indicating a healthy growth state. The CCK-8 assay investigated the quantitative proliferation of the fibroblasts, revealing that after 48 hours of incubation, the relative growth rate value was higher than that of the tissue culture plate, suggesting that the hydrogels do not negatively affect cell adhesion and can even facilitate fibroblast growth. The study concluded that PVA-prGrO-PDA has in vitro biocompatibility, making it suitable for wearable medical devices in portable and real-time health monitoring systems.

A biocompatibility study by Marra et al.(2024), conducted in vitro using a human keratinocyte cell line, HaCaT, indicates that the viability range of the wearable sensors was higher than that of the control sample [66]. It was reported that the integrity of the cell membrane remained intact under various tested conditions.

3.9 Graphene Biodegradability

The biodegradability of graphene and its derivatives is a key factor influencing their behaviour within the body. Biodegradability refers to the ability of microorganisms to change and decompose a material through their metabolic or enzymatic processes. For 2D nanostructures, such as graphene, to gain approval for clinical application and to ensure their safety, it is crucial to demonstrate their complete clearance from the body and their biodegradability. Therefore, understanding the mechanisms that drive biodegradation is closely linked to biocompatibility. Initially, GBMs were considered structurally stable. However, subsequent research has revealed that oxidative enzymes, such as peroxidases, can aid in the breakdown of GrO or carbon nanotubes under laboratory conditions, as well as in vitro and vivo [67–70].

A comparative analysis revealed that the degradation of GrO sheets by hypochlorite occurs more rapidly than that of one-dimensional oxidised carbon nanotubes and nanohorns [71]. It is reported that nanomaterial dispersibility, the synthesis methods employed, and the impact of surface functionalisation affect the biodegradation process [72]. Among GBMs, GrO is the most extensively researched two-dimensional substance in biological applications because of its ability to be easily modified on the surface and dispersed in water. Research into the impact of surface coatings on the biodegradation of GrO and its derivatives demonstrated that both polyethylene glycol (PEG) and bovine serum albumin (BSA) offer protective effects against degradation caused by horseradish peroxidase (HRP) [73].

Research by Kurapati et al. [74] and Mukherjee et al. [75] demonstrated that myeloperoxidase (MPO), a peroxide enzyme released by neutrophil cells responsible for eliminating bacteria and other foreign entities, particularly in the lungs, can effectively biodegrade graphene oxide. However, the structure of non-functionalised graphene was thought to be more resistant to degradation. To investigate this phenomenon, Bianco et al. conducted experiments to determine if and how MPO can break down graphene [72, 74]. They examined the effects of MPO on two forms of graphene—single-layer and few-layer—utilising different fabrication techniques in water. These graphene samples were exposed to MPO in the presence of hydrogen peroxide. Interestingly, non-functionalised graphene was found to be susceptible to degradation. It was found that when bacteria or foreign substances are detected, neutrophils encapsulate them and secrete MPO to neutralise the threat. Their findings indicated that neutrophils can break down highly dispersible graphene within the body, paving the way for developments in graphene and its derivatives [74].

Alternative artificial enzymes, like DNAzymes with a PS2 M-hemin complex that mimics HRP, have also been reported to degrade GrO [76]. However, additional studies to test both in vitro and in vivo degradation and elimination of GBMs are necessary to exclude their possible long-term accumulation and persistence.

3.10 Summary

There is ongoing discussion surrounding the biocompatibility of GBMs, as the available market options and synthesis methods are highly heterogeneous. With only a few studies available, the exact toxicity of GBMs after cutaneous exposure or any conclusions regarding the dermal effects of GBMs are still insufficiently defined. Moreover, not all materials labelled as 'graphene' and graphene-based are identical, as their properties can differ based on physicochemical factors and interaction with biological systems. Direct interaction with the cell membrane may result in varying cell toxicity levels. Gro demonstrates greater suitability for biomedical applications due to its tunable surface chemistry, enhanced solubility, and robust stability within biological fluids. Future research should aim to optimise the functionalisation of GBMs to maximise their efficacy while minimising potential cytotoxicity or other adverse effects. The next chapter presents the application of graphene as a non-invasive biophysical sensor.

References

1. S. Gurunathan, J.-H. Kim, Synthesis, toxicity, biocompatibility, and biomedical applications of graphene and graphene-related materials. Int. J. Nanomed. 1927–1945 (2016)

2. H. Zhang, R. He, Y. Niu, F. Han, J. Li, X. Zhang, F. Xu, Graphene-enabled wearable sensors for healthcare monitoring. Biosens. Bioelectron. **197**, 113777 (2022)
3. I. Manavi-Tehrani, M. Rabiee, M. Parviz, M.R. Tahriri, Z. Fahimi, Preparation, characterization and controlled release investigation of biocompatible ph-sensitive pva/paa hydrogels, in *Macromolecular Symposia*, vol. 296 (Wiley Online Library, 2010), pp. 457–465
4. M. Tahriri, M. Del Monico, A. Moghanian, M.T. Yaraki, R. Torres, A. Yadegari, L. Tayebi, Graphene and its derivatives: opportunities and challenges in dentistry. Mater. Sci. Eng. C **102**, 171–185 (2019)
5. Y. Wang, H. Li, Y. Cheng, Y. Zheng, L. Ruan, In vitro and in vivo studies on ti-based bulk metallic glass as potential dental implant material. Mater. Sci. Eng. C **33**(6), 3489–3497 (2013)
6. R. Guazzo, C. Gardin, G. Bellin, L. Sbricoli, L. Ferroni, F.S. Ludovichetti, A. Piattelli, I. Antoniac, E. Bressan, B. Zavan, Graphene-based nanomaterials for tissue engineering in the dental field. Nanomaterials **8**(5), 349 (2018)
7. A. Ruiz, M.A. Lucherelli, D. Murera, D. Lamon, C. M´enard-Moyon, A. Bianco, Toxicological evaluation of highly water dispersible few-layer graphene in vivo. Carbon **170**, 347–360 (2020)
8. A. Rhazouani, H. Gamrani, M. El Achaby, K. Aziz, L. Gebrati, M.S. Uddin, F. Aziz, Synthesis and toxicity of graphene oxide nanoparticles: a literature review of in vitro and in vivo studies. Biomed. Res. Int. **2021**(1), 5518999 (2021)
9. M. Rasoulianboroujeni, F. Fahimipour, P. Shah, K. Khoshroo, M. Tahriri, H. Eslami, A. Yadegari, E. Dashtimoghadam, L. Tayebi, Development of 3d-printed plga/tio2 nanocomposite scaffolds for bone tissue engineering applications. Mater. Sci. Eng. C **96**, 105–113 (2019)
10. H. Eslami, M. Tahriri, F. Moztarzadeh, R. Bader, L. Tayebi, Nanostructured hydroxyapatite for biomedical applications: from powder to bioceramic. J. Korean Ceram. Soc. **55**(6), 597–607 (2018)
11. A. Shahin-Shamsabadi, A. Hashemi, M. Tahriri, A viscoelastic study of poly (ε-caprolactone) microsphere sintered bone tissue engineering scaffold. J. Med. Biol. Eng. **38**, 359–369 (2018)
12. T. Almela, S. Al-Sahaf, I.M. Brook, K. Khoshroo, M. Rasoulianboroujeni, F. Fahimipour, M. Tahriri, E. Dashtimoghadam, R. Bolt, L. Tayebi et al., 3d printed tissue engineered model for bone invasion of oral cancer. Tissue Cell **52**, 71–77 (2018)
13. F. Fahimipour, E. Dashtimoghadam, M. Rasoulianboroujeni, M. Yazdimamaghani, K. Khoshroo, M. Tahriri, A. Yadegari, J.A. Gonzalez, D. Vashaee, D.C. Lobner et al., Collagenous matrix supported by a 3d-printed scaffold for osteogenic differentiation of dental pulp cells. Dent. Mater. **34**(2), 209–220 (2018)
14. Z. Cheng, M. Li, R. Dey, Y. Chen, Nanomaterials for cancer therapy: current progress and perspectives. J. Hematol. Oncol. **14**, 1–27 (2021)
15. W. Yu, L. Sisi, Y. Haiyan, L. Jie, Progress in the functional modification of graphene/graphene oxide: a review. RSC Adv. **10**(26), 15328–15345 (2020)
16. S. Duan, R. Wu, Y.-H. Xiong, H.-M. Ren, C. Lei, Y.-Q. Zhao, X.-Y. Zhang, F.-J. Xu, Multifunctional antimicrobial materials: from rational design to biomedical applications. Prog. Mater Sci. **125**, 100887 (2022)
17. N. Liu, M. Tang, J. Ding, The interaction between nanoparticles-protein corona complex and cells and its toxic effect on cells. Chemosphere **245**, 125624 (2020)
18. Y. Zhang, S.F. Ali, E. Dervishi, Y. Xu, Z. Li, D. Casciano, A.S. Biris, Cytotoxicity effects of graphene and single-wall carbon nanotubes in neural phaeochromocytoma-derived pc12 cells. ACS Nano **4**(6), 3181–3186 (2010)
19. E. Van Tienhoven, D. Korbee, L. Schipper, H. Verharen, W. De Jong, In vitro and in vivo (cyto) toxicity assays using pvc and ldpe as model materials. J. Biomed. Mater. Res. Part A Official J. Soc. Biomater. Jpn. Soc. Biomater. Aust. Soc. Biomater. Korean Soc. Biomater. **78**(1), 175–182 (2006)
20. Y. Xiao, Y.X. Pang, Y. Yan, P. Qian, H. Zhao, S. Manickam, T. Wu, C.H. Pang, Synthesis and functionalization of graphene materials for biomedical applications: recent advances, challenges, and perspectives. Adv. Sci. **10**(9), 2205292 (2023)
21. S. Mohammadi, A. Babaei, Z. Arab-Bafrani, Polyethylene glycol-decorated go nanosheets as a well-organized nanohybrid to enhance the performance of chitosan biopolymer. J. Polym. Environ. **30**(12), 5130–5147 (2022)

22. Z. Xu, L. Zou, F. Xie, X. Zhang, X. Ou, G. Gao, Biocompatible carboxymethyl chitosan/go-based sponge to improve the efficiency of hemostasis and wound healing. ACS Appl. Mater. Interfaces **14**(39), 44799–44808 (2022)
23. J.L. Patarroyo, J. Cifuentes, L.N. Mun˜oz, J.C. Cruz, L.H. Reyes: Novel antibacterial hydrogels based on gelatin/polyvinyl-alcohol and graphene oxide/silver nanoconjugates: formulation, characterization, and preliminary biocompatibility evaluation. Heliyon **8**(3) (2022)
24. R. Teixeira-Santos, S. Belo, R. Vieira, F.J. Mergulh˜ao, L.C. Gomes, Graphene-based com posites for biomedical applications: Surface modification for enhanced antimicrobial activity and biocompatibility. Biomolecules **13**(11), 1571 (2023)
25. M. Azizi-Lalabadi, H. Hashemi, J. Feng, S.M. Jafari, Carbon nanomaterials against pathogens; the antimicrobial activity of carbon nanotubes, graphene/graphene oxide, fullerenes, and their nanocomposites. Adv. Coll. Interface. Sci. **284**, 102250 (2020)
26. V. Agarwal, P.B. Zetterlund, Strategies for reduction of graphene oxide–a comprehensive review. Chem. Eng. J. **405**, 127018 (2021)
27. K. Gwon, I. Han, S. Lee, Y. Kim, D.N. Lee, Novel metal–organic framework-based photocrosslinked hydrogel system for efficient antibacterial applications. ACS Appl. Mater. Interfaces. **12**(18), 20234–20242 (2020)
28. A. Bolotsky, D. Butler, C. Dong, K. Gerace, N.R. Glavin, C. Muratore, J.A. Robinson, A. Ebrahimi, Two-dimensional materials in biosensing and healthcare: from in vitro diagnostics to optogenetics and beyond. ACS Nano **13**(9), 9781–9810 (2019)
29. N. Rohaizad, C.C. Mayorga-Martinez, M. Fojt, N.M. Latiff, M. Pumera, Two-dimensional materials in biomedical, biosensing and sensing applications. Chem. Soc. Rev. **50**(1), 619–657 (2021)
30. K.-H. Liao, Y.-S. Lin, C.W. Macosko, C.L. Haynes, Cytotoxicity of graphene oxide and graphene in human erythrocytes and skin fibroblasts. ACS Appl. Mater. Interfaces **3**(7), 2607–2615 (2011)
31. A. Sasidharan, L.S. Panchakarla, A.R. Sadanandan, A. Ashokan, P. Chandran, C.M. Girish, D. Menon, S.V. Nair, C. Rao, M. Koyakutty, Hemocompatibility and macrophage response of pristine and functionalized graphene. Small **8**(8), 1251–1263 (2012)
32. C. Yang, X. Hou, L. Zhang, Microfluidics-derived microfibers in flexible bioelectronics. Mater. Futures **3**(3), 032401 (2024)
33. M. Asaftei, M. Lucidi, S.R. Anton, A.-F. Trompeta, R. Hristu, D.E. Tranca, E. Fiorentis, C. Cirtoaje, V. Lazar, G.A. Stanciu et al., Antibacterial interactions of ethanol-dispersed multi-walled carbon nanotubes with staphylococcus aureus and pseudomonas aeruginosa. ACS Omega **9**(31), 33751–33764 (2024)
34. P.A. Hall, F.M. Watt, Stem cells: the generation and maintenance of cellular diversity. Development **106**(4), 619–633 (1989)
35. J. Wang, Y. Tian, X. Shi, Z. Feng, L. Jiang, Y. Hao, Safety and efficacy of cell transplantation on improving motor symptoms in patients with Parkinson's disease: a meta-analysis. Front. Hum. Neurosci. **16**, 849069 (2022)
36. M.T. Insights, Global Guide: Stem Cells for Stroke. https://www.insights.medicaltourism.com/article/global-guide-stem-cells-stroke. Accessed 27 March 2025
37. M. Kalbacova, A. Broz, J. Kong, M. Kalbac, Graphene substrates promote adherence of human osteoblasts and mesenchymal stromal cells. Carbon **48**(15), 4323–4329 (2010)
38. S.D. Newby, T. Masi, C.D. Griffin, W.J. King, A. Chipman, S. Stephenson, D.E. Anderson, A.S. Biris, S.E. Bourdo, M. Dhar, Functionalized graphene nanoparticles induce human mesenchymal stem cells to express distinct extracellular matrix proteins mediating osteogenesis. International journal of nanomedicine, pp. 2501–2513 (2020)
39. N.F. Nicoletti, D.R. Marinowic, D. Perondi, J.I. Budelon Gon¸calves, D. Piazza, J.C. Costa, A. Falavigna, Non-cytotoxic graphene nanoplatelets upregulate cell proliferation and self- renewal genes of mesenchymal stem cells. Int. J. Mol. Sci. **25**(18), 9817 (2024)
40. A.J. Engler, S. Sen, H.L. Sweeney, D.E. Discher, Matrix elasticity directs stem cell lineage specification. Cell **126**(4), 677–689 (2006)

41. N. Jaiswal, S.E. Haynesworth, A.I. Caplan, S.P. Bruder, Osteogenic differentiation of purified, culture-expanded human mesenchymal stem cells in vitro. J. Cell. Biochem. **64**(2), 295–312 (1997)
42. T.R. Nayak, H. Andersen, V.S. Makam, C. Khaw, S. Bae, X. Xu, P.-L.R. Ee, J.-H. Ahn, B.H. Hong, G. Pastorin et al., Graphene for controlled and accelerated osteogenic differentiation of human mesenchymal stem cells. ACS Nano **5**(6), 4670–4678 (2011)
43. W.C. Lee, C.H.Y. Lim, H. Shi, L.A. Tang, Y. Wang, C.T. Lim, K.P. Loh, Origin of enhanced stem cell growth and differentiation on graphene and graphene oxide. ACS Nano **5**(9), 7334–7341 (2011)
44. W.C.L. Kenry, K.P. Loh, C.T. Lim, When stem cells meet graphene: opportunities and challenges in regenerative medicine. Biomaterials **155**, 236–250 (2018)
45. S.Y. Park, J. Park, S.H. Sim, M.G. Sung, K.S. Kim, B.H. Hong, S. Hong, Enhanced differentiation of human neural stem cells into neurons on graphene. Adv. Mater. **23**(36), 263 (2011)
46. R. Guo, S. Zhang, M. Xiao, F. Qian, Z. He, D. Li, X. Zhang, H. Li, X. Yang, M. Wang et al., Accelerating bioelectric functional development of neural stem cells by graphene coupling: implications for neural interfacing with conductive materials. Biomaterials **106**, 193–204 (2016)
47. G.-Y. Chen, D.-P. Pang, S.-M. Hwang, H.-Y. Tuan, Y.-C. Hu, A graphene-based platform for induced pluripotent stem cells culture and differentiation. Biomaterials **33**(2), 418–427 (2012)
48. K. Takahashi, S. Yamanaka, Induction of pluripotent stem cells from mouse embryonic and adult fibroblast cultures by defined factors. Cell **126**(4), 663–676 (2006)
49. W. Hu, C. Peng, W. Luo, M. Lv, X. Li, D. Li, Q. Huang, C. Fan, Graphene-based antibacterial paper. ACS Nano **4**(7), 4317–4323 (2010)
50. S. Liu, T.H. Zeng, M. Hofmann, E. Burcombe, J. Wei, R. Jiang, J. Kong, Y. Chen, Antibacterial activity of graphite, graphite oxide, graphene oxide, and reduced graphene oxide: membrane and oxidative stress. ACS Nano **5**(9), 6971–6980 (2011)
51. V.T. Pham, V.K. Truong, M.D. Quinn, S.M. Notley, Y. Guo, V.A. Baulin, M. Al Kobaisi, R.J. Crawford, E.P. Ivanova, Graphene induces formation of pores that kill spherical and rod-shaped bacteria. ACS Nano **9**(8), 8458–8467 (2015)
52. O. Akhavan, E. Ghaderi, A. Esfandiar, Wrapping bacteria by graphene nanosheets for isolation from environment, reactivation by sonication, and inactivation by near-infrared irradiation. J. Phys. Chem. B **115**(19), 6279–6288 (2011)
53. R. Zhou, H. Gao, Cytotoxicity of graphene: recent advances and future perspective. Wiley Interdiscip. Rev. Nanomed. Nanobiotechnol. **6**(5), 452–474 (2014)
54. S. Liu, M. Hu, T.H. Zeng, R. Wu, R. Jiang, J. Wei, L. Wang, J. Kong, Y. Chen, Lateral dimension-dependent antibacterial activity of graphene oxide sheets. Langmuir **28**(33), 12364–12372 (2012)
55. J. He, X. Zhu, Z. Qi, C. Wang, X. Mao, C. Zhu, Z. He, M. Li, Z. Tang, Killing dental pathogens using antibacterial graphene oxide. ACS Appl. Mater. Interfaces **7**(9), 5605–5611 (2015)
56. E. Karatan, P. Watnick, Signals, regulatory networks, and materials that build and break bacterial biofilms. Microbiol. Mol. Biol. Rev. **73**(2), 310–347 (2009)
57. C. Song, C.-M. Yang, X.-F. Sun, P.-F. Xia, J. Qin, B.-B. Guo, S.-G. Wang, Influences of graphene oxide on biofilm formation of gram-negative and gram-positive bacteria. Environ. Sci. Pollut. Res. **25**, 2853–2860 (2018)
58. P.-C. Wu, H.-H. Chen, S.-Y. Chen, W.-L. Wang, K.-L. Yang, C.-H. Huang, H.-F. Kao, J.-C. Chang, C.-L.L. Hsu, J.-Y. Wang et al., Graphene oxide conjugated with polymers: a study of culture condition to determine whether a bacterial growth stimulant or an antimicrobial agent? J. Nanobiotechnol. **16**, 1–20 (2018)
59. M. Kong, M. Yang, R. Li, Y.-Z. Long, J. Zhang, X. Huang, X. Cui, Y. Zhang, Z. Said, C. Li, Graphene-based flexible wearable sensors: mechanisms, challenges, and future directions. Int. J. Adv. Manuf. Technol. **131**(5), 3205–3237 (2024)
60. A. Nag, R.B. Simorangkir, D.R. Gawade, S. Nuthalapati, J.L. Buckley, B. O'Flynn, M.E. Altinsoy, S.C. Mukhopadhyay, Graphene-based wearable temperature sensors: a review. Mater. Des. **221**, 110971 (2022)

61. M. Pelin, S. Sosa, M. Prato, A. Tubaro, Occupational exposure to graphene based nanomaterials: risk assessment. Nanoscale **10**(34), 15894–15903 (2018)
62. C. Yin, L. Yu, L. Feng, J.T. Zhou, C. Du, X. Shao, Y. Cheng, Nanotoxicity of two- dimensional nanomaterials on human skin and the structural evolution of keratin protein. Nanotechnology **35**(22), 225101 (2024)
63. B. Fadeel, C. Bussy, S. Merino, E. V´azquez, E. Flahaut, F. Mouchet, L. Evariste, L. Gauthier, A.J. Koivisto, U. Vogel, et al., Safety assessment of graphene-based materials: focus on human health and the environment. ACS Nano **12**(11), 10582–10620 (2018)
64. X. Zhang, Y. Wang, G. Luo, M. Xing, Two-dimensional graphene family material: assembly, biocompatibility and sensors applications. Sensors **19**(13), 2966 (2019)
65. Y. Zhang, B. Liang, Q. Jiang, Y. Li, Y. Feng, L. Zhang, Y. Zhao, X. Xiong, Flexible and wearable sensor based on graphene nanocomposite hydrogels. Smart Mater. Struct. **29**(7), 075027 (2020)
66. F. Marra, A. Preziosi, A. Tamburrano, C.J. Kundukulam, P. Mancini, D. Uccelletti, M.S. Sarto, Study, design and development of biocompatible graphene-based piezoresistive wearable sensors for human monitoring. IEEE Sens. J. **24**(5), 6709–6718 (2024)
67. J. Russier, L. Oudjedi, M. Piponnier, C. Bussy, M. Prato, K. Kostarelos, B. Lounis, A. Bianco, L. Cognet, Direct visualization of carbon nanotube degradation in primary cells by photothermal imaging. Nanoscale **9**(14), 4642–4645 (2017)
68. G.P. Kotchey, B.L. Allen, H. Vedala, N. Yanamala, A.A. Kapralov, Y.Y. Tyurina, J. Klein-Seetharaman, V.E. Kagan, A. Star, The enzymatic oxidation of graphene oxide. ACS Nano **5**(3), 2098–2108 (2011)
69. G.P. Kotchey, S.A. Hasan, A.A. Kapralov, S.H. Ha, K. Kim, A.A. Shvedova, V.E. Kagan, A. Star, A natural vanishing act: the enzyme-catalyzed degradation of carbon nanomaterials. Acc. Chem. Res. **45**(10), 1770–1781 (2012)
70. K. Bhattacharya, S.P. Mukherjee, A. Gallud, S.C. Burkert, S. Bistarelli, S. Bellucci, M. Bottini, A. Star, B. Fadeel, Biological interactions of carbon-based nanomaterials: from coronation to degradation. Nanomed. Nanotechnol. Biol. Med. **12**(2), 333–351 (2016)
71. L. Newman, N. Lozano, M. Zhang, S. Iijima, M. Yudasaka, C. Bussy, K. Kostarelos, Hypochlorite degrades 2d graphene oxide sheets faster than 1d oxidised carbon nanotubes and nanohorns. npj 2D Mater. Appl. **1**(1), 39 (2017)
72. R. Kurapati, J. Russier, MA. Squillaci, E. Treossi, C. M´enard-Moyon, A.E. Del Rio-Castillo, E. Vazquez, P. Samor'ı, V. Palermo, A. Bianco, Dispersibility-dependent biodegradation of graphene oxide by myeloperoxidase. Small **11**(32), 3985–3994 (2015)
73. Y. Li, L. Feng, X. Shi, X. Wang, Y. Yang, K. Yang, T. Liu, G. Yang, Z. Liu, Surface coating-dependent cytotoxicity and degradation of graphene derivatives: towards the design of non-toxic, degradable nano-graphene. Small **10**(8), 1544–1554 (2014)
74. R. Kurapati, S.P. Mukherjee, C. Mart´ın, G. Bepete, E. V´azquez, A. P´enicaud, B. Fadeel, A. Bianco, Degradation of single-layer and few-layer graphene by neutrophil myeloperoxidase. Angewandte Chemie Int. Edn. **57**(36), 11722–11727 (2018)
75. S.P. Mukherjee, A.R. Gliga, B. Lazzaretto, B. Brandner, M. Fielden, C. Vogt, L. Newman, A.F. Rodrigues, W. Shao, P.M. Fournier et al., Graphene oxide is degraded by neutrophils and the degradation products are non-genotoxic. Nanoscale **10**(3), 1180–1188 (2018)
76. R. Kurapati, A. Bianco, Peroxidase mimicking dnazymes degrade graphene oxide. Nanoscale **10**(41), 19316–19321 (2018)

Chapter 4
Application of Graphene as a Non-invasive Biophysical Sensor

Abstract This chapter examines the application of graphene and its derivatives as non-invasive biophysical sensors, highlighting their potential in monitoring various health parameters owing to their superior electronic and mechanical properties. The versatility of graphene enables continuous, real-time observation of minor fluctuations in electrophysiological signals, such as electrocardiograms (ECG), electromyograms (EMG), electrooculograms (EOG), and electroencephalograms (EEG), alongside kinematic signals like pulse wave, joint movement, respiratory rate, and body temperature. This capability represents an improvement in personal healthcare, particularly in the early identification of conditions such as hypertension, arthritis, and respiratory disorders. The intrinsic nature of electrophysiological signals, generated by active cells through the movement of ions across membranes, resulting in resting and action potentials, is also presented. The accurate analysis of these signals is often negatively affected by technical challenges such as noise, weak signal strength, and high contact impedance. Consequently, the report presents the importance of electrode quality in bioelectrical signal measurement. It was suggested that optimal bioelectrical sensors should exhibit high accuracy, a robust signal-to-noise ratio (SNR), low impedance, a broad dynamic range, and exceptional durability. The synthesis of these characteristics in graphene-based electrode sensors positions them at the forefront of biomedical technology, potentially advancing diagnostic methodologies and patient monitoring systems in clinical settings.

Keywords Graphene · Biophysical sensors · Electrophysiological signals · Non-invasive monitoring · Wearable technology · Electrocardiogram (ECG) · Electromyography (EMG) · Electroencephalogram (EEG) · Motion · Pressure · Kinematic · Acoustic sensor · Temperature

S. Debnath et al., *Graphene in Wearable Sensors for Health Monitoring*,
SpringerBriefs in Applied Sciences and Technology,
https://doi.org/10.1007/978-981-96-8850-0_4

4.1 Introduction

Traditional health monitoring and diagnosis methods generally require large equipment, invasive sampling techniques such as blood drawing, and bench-top procedures that can cause discomfort and pain for patients [1]. Moreover, it can also be time-consuming and inconvenient. However, wearable biomedical sensors have gained interest in the market. Human health monitoring depends on physical, chemical, and biological information transmitted through the skin [2, 3]. Wearable sensors can detect physical, chemical, and biological information accurately and in real-time [4–6] by attaching to various body parts as an analytical tool [7–9]. These data are used to diagnose and treat different health conditions [10–13]. In biomedical applications, wearable sensors are recognised for their ability to be well-designed, their excellent sensing capabilities for precise and stable biosignal detection, and their optimal mechanical flexibility for conformable integration with human bodies [14, 15].

Due to the inconsistency between human skin and conventional rigid silicon-based sensors, sensors should exhibit mechanical flexibility for optimal performance [16]. Factors such as biocompatibility, reliability, stability, comfort, convenience, miniaturisation, cost, and biofouling must also be considered to ensure long-term, multi-functional, real-time, affordable, and unobtrusive health monitoring without limitations [17]. Therefore, selecting sensing materials, the core element of biomedical sensors, is crucial. These materials should possess characteristics such as responsiveness to biophysical or biochemical stimuli for the reliable sensing of health-related signals and compatibility with mechanical structure design for improved tolerance to deformations during applications [14]. Graphene and its derivatives hold promising potential for monitoring biophysical parameters due to their exceptional electronic and mechanical properties. Their applications are visible, ranging from personal healthcare to human-machine bio-interfaces [18].

Graphene has several advantages when used in wearable sensors [17]. Firstly, due to its high specific surface area and atomic thickness, graphene layers allow entire carbon atoms to come into direct contact with analytes, resulting in sensors with superior sensitivity compared to silicon [19–22]. Additionally, due to the mechanical flexibility and ultrathin thickness of graphene, graphene- based sensors are adaptive and present intense contact with organs, such as the skin [23, 24] and eyes [25–27]. These features are essential for acquiring high-quality signals without irritation, motion artefacts, or contamination [28]. Graphene's high optical transparency and electrical conductivity are ideal for observing bio-tissue with clear images and minimal visual disturbances [29]. Furthermore, due to its high electrical conductivity, conformal integration and efficient signal transmission enable a high signal-to-noise ratio (SNR) in electrophysiological signal recording. For instance, minor changes in electrophysiological signals like Electrocardiogram (ECG), Electromyography (EMG), Electrooculography (EOG), and Electroencephalogram (EEG) and kinematic signals such as pulse wave, joint movement, respiratory rate, and body temperature can be continuously observed in real-time. This capability offers the potential for early warning signs of conditions such as hypertension, arthritis, and respiratory diseases [30].

4.2 Electrophysiological

Humans and other living organisms within active cells naturally generate electrophysiological signals. These signals originate from the movement of ions across cell membranes, resulting in both resting and action potentials. Resting potentials occur due to an uneven distribution of sodium and potassium ions, which creates a potential difference between the inside and outside of the cell. The active potential is triggered by external stimulation and causes transient changes in the membrane potential [17]. Electrodes in ECG, EEG, EMG, EOG, and other diagnostic tools are used to detect and analyse electrophysiological signals [17, 31]. These signals are crucial in preventing and treating various diseases [32]. However, obtaining accurate signal analysis poses a challenge due to common- mode noise, weak signal, and high contact impedance. Therefore, the quality of electrodes is critical in measuring the bioelectrical signal. Bioelectrical electrode sensors should possess accuracy, a high signal-to-noise ratio (SNR), low impedance, a broad dynamic range, robustness, durability, and excellent repeatability across a wide range of deformations. However, current bio-electrodes face challenges, such as expensive manufacturing and poor skin contact, resulting in an unstable signal acquisition in kinetic states [17]. Silver/silver chloride (Ag/AgCl) electrodes are widely used to record bioelectricity from different physical activities [33]. However, the gel and adhesives used in these wet electrodes may cause skin irritation and allergic reactions, making them unsuitable for extended and wearable applications [34]. These limitations hinder the collection of real-time bioelectrical signals. With the development of wearable biometric sensors, innovative solutions exist for real-time, high-fidelity, low-impedance, and SNR bioelectrical signal acquisition [17]. Dry electrodes, such as graphene-based electrodes, are a promising alternative for long-term monitoring of bioelectricity. They are highly compatible with the body and maintain good skin contact even during movement [31].

4.2.1 Electrocardiography (ECG)

ECG tracks the heart's activity during each cardiac cycle, pacemaker usage, atrial and ventricular excitation, and changes in bioelectricity. By analysing the different shapes of the electrocardiogram, doctors determine whether a patient has any heart-related issues [37].

In 2016, Kim T. et al. developed dry adhesives (CDAs) by incorporating carbon-based materials into a PDMS matrix [38]. This integration established a conductive network within the PDMS elastomer, enhancing its suitability for electronic interface applications. The fabrication process is further refined by utilising an inverse moulding technique, whereby the CDA pad is shaped using a silicon template crafted via photolithography and deep reactive ion etching methods. The resultant CDA-based ECG electrodes were equipped with mushroom-shaped microstructures, featuring an aspect ratio exceeding three, which considerably expanded the contact area. It was accurately reproduced without defects and delivered consistent cyclic adhesion capabilities (approximately 30 cycles) with a normal adhesion force of around 1.3 N cm^{-2} when applied to human skin. This level of performance was comparable to that of existing wet adhesives. To enhance electrical connectivity and maximise conductivity within the elastomer matrix at a minimal filler concentration (approximately 1 weight%), a hybrid filler system that combines 1D and 2D carbon materials was employed, with an optimal blend ratio of 9:1 (CNTs to graphene). Under these optimised conditions, the composite film demonstrated stretchability of over 100% and the conductivity of 0.01 $S.cm^{-1}$, making it suitable for powering LEDs. Additionally, distinct from conventional ECG electrodes, the CDA electrode features advanced properties, including water and dust resistance, as well as the potential for multiple reuse. The high-aspect-ratio micropillars endow the electrode's surface with properties similar to those of a gecko's skin, contributing to its superhydrophobicity and facilitating self-cleaning.

Furthermore, Kim et al. (2019) developed an ECG electrode characterised by its ability to drain water and permit airflow [39]. The conceptual foundation for this innovation was derived from the microchannel networks observed in the toe pads of tree frogs and the convex cup-like structures inherent to octopus suckers. To realise this design, silicon moulds engraved with micro-hole patterns and structures reminiscent of those found in frogs and octopuses were replicated using PDMS. These PDMS constructs were subsequently coated with a layer of rGrO nanoplatelets using a spray coating technique, resulting in a final thickness of approximately 1.2 μm. Further inspired by the ability of rain frogs to adhere to uneven surfaces by directing water through hexagonal channels within their pads, the electrodes were engineered with a hierarchical architecture, enhancing their capacity to maintain contact on moist surfaces through elevated peeling energy. Moreover, to increase the electrode's adhesion on wet surfaces, cylindrical microholes inspired by the protruding structures observed on the surface of octopus suckers were stamped and moulded onto the electrode's contact areas. This microstructure enabled the electrode patches to achieve adhesion strengths of 6.6 $N{\cdot}cm^{-2}$, 5.3 $N{\cdot}cm^{-2}$, and 4.5 $N{\cdot}cm^{-2}$ on dry surfaces, wet surfaces, and under aquatic conditions, respectively. After applying a conductive layer of ultrathin rGrO nanoplates, the completed ECG electrode demonstrated effectiveness in monitoring electrocardiographic signals on wet skin during physical exertion or in aquatic environments.

In 2020, Qiu et al. introduced a durable graphene-based skin electrode for continuous ECG monitoring, drawing inspiration from the solidification process of bird nests [40]. The approach produced the nest's structure and creation process by utilising electrospun polymer nanofibers—comprising Copper(II) Chloride ($CuCl_2$), tyrene-ethylene-butylene-styrene (SEBS), and phenolic resin captured on a substrate coated with a CVD monolayer graphene film. After annealing, the nanofibers firmly adhered to the graphene layer through π-π interactions between the phenolic side groups and metal ions, facilitating the development of a crystalline graphene framework. The entire assembly of graphene and polymer nanofibers was then encased in a SEBS solution to craft the electrodes. These electrodes featured a bird's nest-like configuration, a conductive network composed of graphene and polymer nanofibers, and semi-encased SEBS to enhance mechanical durability. In ECG electrode applications, they demonstrated sheet resistance of approximately 150 $\Omega\ sq^{-1}$ and a SNR of 30 dB. Huang et al. (2021) developed a nanofiber carbon electrode (CB/rGO) composed of carbon nanofibers and rGro for smart clothing to monitor ECG. Initially, a carbon film collector was created using carbon black (CB) and rGrO uniformly dispersed with the help of the SMA-M1000 copolymer. The dispersion level was controlled by varying the amount of SMA-M1000 in the mix. Nanofibers (PVDF/PEDOT/PSS) were developed by utilising a mixture of polyvinylidene difluoride (PVDF) and poly (3,4-ethylene dioxythiophene) polystyrene sulfonate (PEDOT: PSS) nanofibers and deposited on CB/rGO through electrospinning [35]. Together with the nanofiber layer and CB/rGO film, a carbon nanofiber electrode, CB/rGrO/SMA-M1000, was developed. These carbon composite electrodes were reported to facilitate ECG signal detection when integrated into smart textiles. The contact impedance of CB/rGrO/SMA-M1000 samples prepared with different component ratios was 4.6 kΩ (1:1:0), 0.19 kΩ (0:10:1), 0.38 kΩ (5:5:1), 4.6 kΩ (1:1:1), and 61 kΩ (1:1:5). Moreover, it was reported that ECG readings taken with the carbon nanofiber electrodes under movement were more stable than those recorded with traditional Ag/AgCl electrodes. The study highlighted the sheet resistance of 2.5 10^1 $\Omega\ sq^{-1}$, mechanical durability (maintaining a stable circuit after 3000 cycles of use), and hydrophobic nature (water contact angle of 146°) of the carbon nanofiber electrodes. Compared to conventional wet Ag/AgCl electrodes, the developed carbon nanofiber electrodes demonstrated superior skin compatibility and durability [35].

Xu et al. (2020) developed a graphene-coated textile electrode for ECG monitoring to mitigate the long-standing challenges associated with skin allergies and the desiccating impact of gel utilised in ECG monitoring in 2020 [41]. In their research, electrodes were fabricated by applying a screen-printed graphene ink to a pre-modified textile, characterised by a mass per unit area of 130 g m^{-2} and fibre diameters ranging from 10 to 20 μm. In the initial phase of the fabrication process, the surface roughness of the textile was enhanced by applying a polymer layer, achieved via a heat transfer method at a temperature of 180 °C for 10 seconds. This treatment was used to promote adequate adhesion between the cotton fibres and graphene nanoplates. Screen printing technology was then employed to securely attach the conductive graphene patterns to the textile substrate, thereby increasing the durability and washability of the electrodes for prolonged use. These textile-based electrodes demonstrated a resistance of approximately 104 $\Omega\ sq^{-1}$ and could capture ECG signals with a fidelity comparable to traditional silver/silver chloride (Ag/AgCl) electrodes. Moreover, using a cotton textile substrate ensured the maintenance of a high-quality ECG signal acquisition even after repeated washing cycles.

In 2019, Zhang et al. combined silk with graphene, renowned as the thinnest known electrical conductor, to develop a novel ECG electrode, [42] distinguished by its exceptional softness and resemblance to the texture and feel of a tattoo. The fabrication begins with extracting a silk fibroin (SF) solution from silkworm cocoons. Calcium ions (Ca^{2+}) were introduced into the SF solution to enhance the self-healing capacity of the electrodes. This incorporation resulted in the formation of an SF/Ca^{2+} composite rich in hydrogen and coordination bonds, both of which play a key role in the self-repairing mechanism of the ECG electrode. The next phase involved mixing graphene nanoplates into this mixture, producing a Gr/SF/Ca^{2+} suspension. Subsequently, this blend was applied onto the SF/Ca^{2+} base via mask printing or direct writing techniques to develop a silk-based ECG electrode that adhered to the skin like a tattoo.

Furthermore, in 2020, Pan et al. (2020) developed a bionic electrode using conductive hydrogels with a nerve-like nanostructure [43]. This electrode was reported to have self-healing capability, stretchability, and flexibility. By incorporating proanthocyanins/rGO (PC/rGO) into glycerol-plasticised PVA–borax hydrogel framework, they created a PC/rGO/PVA hydrogel that exhibited stretchability (over 5000%), compliance (1 mm), and rapid self-healing capabilities (3 seconds, with 95.73% efficiency). The self-healing property of the hydrogel electrode was primarily attributed to glycerol, which provided numerous free borax-chelating sites for self-repair, mimicking the regenerative abilities of neural networks. The catechol groups on the PC/rGo surface participate in multiple cross-linking interactions with borax, enabling the rapid healing of the hydrogel electrode. Without any baseline fluctuation, this electrode could maintain a high signal-to-noise ratio and record the ECG signals on human skin.

Prasad et al. developed an electrode for real-time ECG signal capture by incorporating a GNR into a PVA matrix, referred to as PVA/GNR [44]. The electrode was created by integrating various amounts of GNR into the PVA polymer and coating the surface of an Ag/AgCl electrode. They achieved higher electrical and mechanical properties in the GNRs by chemically unzipping MWCNTs through oxidation and reduction processes. The biocompatible and water-soluble PVA was combined with the synthesised GNRs to produce the composite material, which was then used to manufacture the ECG electrode. This GNR electrode maintained a consistent impedance of 250 KΩ across a frequency range of 10 Hz to 1 KHz. Notably, the electrode demonstrated excellent performance, including an SNR of 59.99 dB and a 348.62 mV value for the R-peak, with only 1 wt% GNR content, suggesting the potential of using GNR to develop cost-effective graphene-based ECG electrodes. Furthermore, a week-long wearability test confirmed the electrode's ability, indicating no degradation in the ECG signal quality or skin irritation.

Huang et al. (2022) developed a composite electrode for ECG applications comprising copper nanoparticles (CuNPs), polythiophene (PTT), and GrO [45]. This composite was engineered to achieve enhanced air stability and uniform dispersion of the constituents. The PTT was used as a stabiliser for the CuNPs and was attached to the nanoparticles through van der Waals forces and ionic charge interactions. A key characteristic of this composite was the presence of epoxy groups in GrO, which promoted the attraction of Cu^{2+} ions, thus enhancing the nanoscale stability of the copper nanoparticles (CuNPs). The study's findings indicated that the CuNPs/

PTT/GrO composite film retained a conductive network with a sheet resistance of 9.06 10^{-1} Ω sq^{-1}, even after 30 days of storage. Consequently, the CuNPs/PTT/GrO composite film was integrated into ECG smart clothing, yielding a durable and oxidation-resistant sensing electrode.

Taking into account the sliding phenomenon between graphene layers while preserving the integrity of interlayer connections, Du et al. (2020) proposed a graphene-based ECG electrode through the utilisation of bilayer graphene intercalated with molybdenum chloride ($MoCl_5$), hence referred to as Mo-BLG [46]. CVD synthesised single-layer graphene (SLG) on Cu foil, and a standard wet transfer method was used for layer-by-layer stacking of graphene. This structured multilayer graphene was subsequently positioned within a tube containing a nitrogen atmosphere, in conjunction with $MoCl_5$ and MoO_3 powder, and subjected to thermal treatment afterwards. The resultant Mo-BLG was then transitioned onto a SEBS substrate and encapsulated. It was reported that the charge transfer between the dopants and the graphene layers reduced the sheet resistance from 220 to 40 Ω sq^{-1} while achieving transparency of over 80%. In the sandwich structure, the nanoconfined doping effect slightly increased the sheet resistance of Mo-BLG to 60 Ω sq^{-1} after 30 days. Furthermore, Mo-BLG applied to an elastomer remained conductive under a tensile strain of up to 80%, consistently maintaining its conductivity over 2000 cycles at 30%.

Suvarnaphaet et al. (2019) investigated a biodegradable biomedical electrode patch to capture ECG signals [47]. This patch featured a two-part design that included a chemically produced graphene layer for conductivity and polyhydroxyalkanoate (PHA), a flexible material, creating an effective Gr/PHA blend. This blend was anticipated to be fully biocompatible with human tissues and capable of breaking down biologically in soil environments. Hummers' method [48] and a hydrazine hydrate reduction step were used to synthesise graphene. At the same time, the PHA layer was created through the microbial fermentation process involving Ralstonia eutropha. This layer was then turned into a nanofiber scaffold using electrospinning, setting the stage for the integration of graphene. The resulting Gr/PHA electrodes were connected to a computational microcontroller, which managed the gathering and amplification of data from a 3-lead ECG sensor [47]. The performance of these electrodes was evaluated through in vitro simulations of ECG signals, encompassing both normal and arrhythmic heart rhythms. These tests showed that the Gr/PHA electrodes could maintain signal quality over time. However, it was theorised that signal weakening might occur during use in a live body due to the resistance encountered at the interface between the human skin and the electrode [47]. Notably, although the electrode material was inherently biodegradable, it was found that the electrodes could last for 1–2 years, depending on the environmental conditions, particularly moisture and oxygen levels.

Murastov et al. (2020) demonstrated the creation of bioelectrodes using laser-rGrO on flexible polyethylene terephthalate (PET) substrates [36]. This development aims to address two challenges in the field: the use of hydrogels in skin interfaces with traditional Ag/AgCl bioelectrodes and the historically low SNR found in capacitance or dry electrodes. Their dry rGrO bioelectrode technology was presented to maintain stability for up to 100 hours in harsh conditions and upon skin contact. Through rigorous reliability tests across various buffer solutions with pH levels ranging from 4.8 to 9.2 over 24 hours, the rGrO electrodes demonstrated exceptional durability. Regarding SNR, these bioelectrodes presented an SNR of over 98%, with a signal match comparable to that of leading-edge electrodes used for comparison purposes. The unparalleled stability of the bioelectrodes can be attributed to the modifications and composite formation at the rGrO/PET interface resulting from the laser process applied for GrO reduction.

Cui et al. (2022) proposed a flexible ECG sensor utilising a combination of graphene and PVA film via a double transfer method [49]. In this design, graphene was the active layer, while the PVA film provided a flexible base for the electrode. The sensor was reported to have Young's modulus of 8.598 MPa, a sheet resistance of 75 $\Omega\ sq^{-1}$ and a contact impedance of 45.12 kΩ at 100 Hz. Moreover, the sensor could maintain resistance within a 20% strain range and detect ECG signals with an SNR of 60 dB. This flexible ECG sensor may also undergo delamination within 150 seconds upon application of deionised water.

Zahed et al. (2021) introduced a flexible 3D porous graphene (3DPG) electrode incorporating polyaziridine-encapsulated phosphorene (PEP) using cost-effective laser writing and drop-casting techniques [50]. The PEP/3DPG carcinoembryonic immunosensor was reported to present linear ranges of 0.1–700 pg mL^{-1} and 1–100 ng mL^{-1}, with a detection limit of 0.34 pg mL^{-1} and high selectivity under optimised conditions. The ECG sensor based on the PEP/3DPG electrode exhibited stable impedance at the skin-electrode interface, resulting in an SNR of 13.5 dB, comparable to that of conventional Ag/AgCl electrodes (13.9 dB).

Zhang et al. (2022) developed a stretchable ECG electrode using LIG techniques, incorporating ultrathin and highly flexible polystyrene-block-poly(ethylene-ran-butylene)-block-polystyrene (SEBS) into a stretchable, multi-layer structure [51]. Initially, LIG was produced on a PI film employing a CO_2 laser that operated at a wavelength of 10.6 μm. Subsequently, 3,4-ethylenedioxythiophene (EDOT) was combined with MXene-Ti_3C_2 to enhance the LIG electrode's temperature and strain sensing capabilities, and improve its electrical conductivity and compatibility with SEBS, facilitating more effortless LIG transfer and encapsulation. An MXene-Ti_3C_2@EDOT layer was added to the LIG via electrophoretic deposition. This MXene-LIG composite was then encapsulated from the PI film and sent to the SEBS. This innovation resulted in highly flexible and conductive electrodes capable of detecting ECG signals, monitoring temperature, and tracking movement. Compared to traditional Ag/AgCl gel electrodes, the MXene-LIG electrode exhibited a lower contact impedance (51.08 kΩ at 10 Hz). Its stability and flexibility also contributed to a higher SNR ratio for continuous and real-time ECG monitoring, marking an advancement in the development of graphene-based ECG electrodes.

Maher et al. developed an ECG electrode using the esterification process to combine glycerol (Gl) with a PVA polymer base [52]. Initially, they synthesised GrO using a modified Hummers method and mixed it with Gl in an acidic environment (pH 1–3). Once the reaction was complete, the conductive mixture of GrO and Gl was blended with PVA to create a flexible conductor. This composite of GrO, Gl,

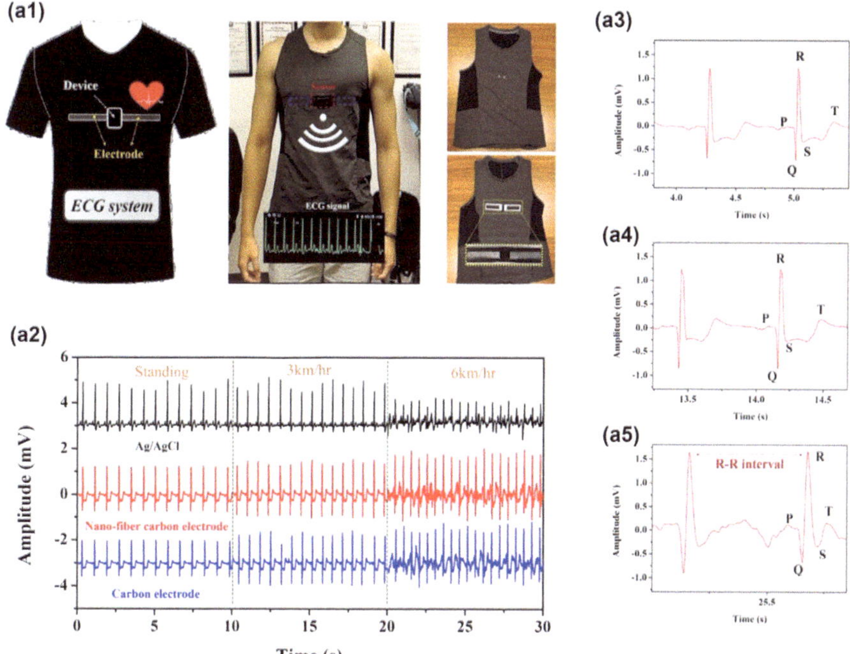

Fig. 4.1 Measurement and analysis of ECG signals. **a1** Design of the nanofiber carbon electrode on the clothing. **a2** Comparison between the ECG signals collected by the three sensor types during exercise. **a3** Peak values of the P, Q-R-S, and T in the ECG signal after amplification at t = 4.2 s. **a4** Amplified signal at t = 14.2 s. (b5) Amplified signal at t = 25.6 s. (Reproduced from [35] with permission from ACS Publications)

and PVA was then moulded into the shape of an ECG electrode. Using PVA as a carrier for GrO/Gl enabled an even distribution of conductive, reduced GrO sheets. Additionally, due to PVA's water solubility, the electrodes could be applied to the skin with water, ensuring flexibility, good adhesion, and low impedance with human skin. This results in ECG electrodes that effectively capture signals with a high SNR.

Figures 4.1, 4.2 and Table 4.1 present different types of ECG sensors and their features.

4.2.2 *Electromyography (EMG)*

Electromyography (EMG) records the electrical activity produced by skeletal muscles, both at rest and during muscular contraction. It is achieved through electronic instruments that examine the excitability and conduction functions of nerves

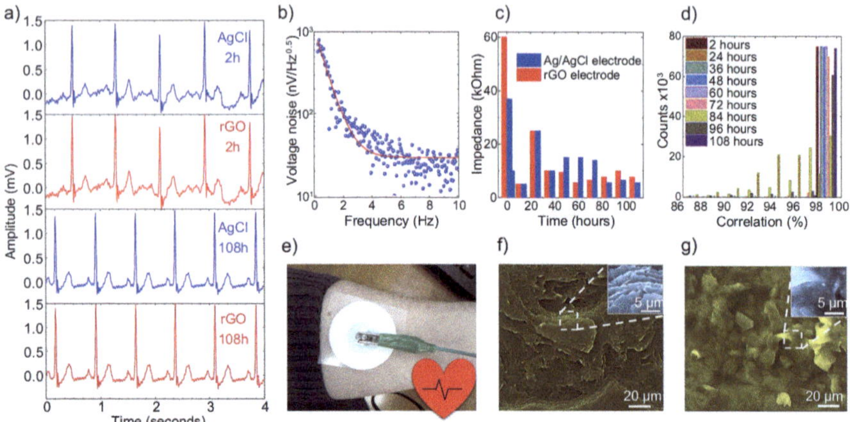

Fig. 4.2 ECG processed data: **a** The ECG signals with rGO bioelectrode compared to the Ag/AgCl electrode at different measurement times; **b** Voltage noise vs. frequency; **c** The bioimpedance change of skin–electrode contact during the test; **d** Signal correlation from different electrodes with 5 min ECG data at different times (over 98% match obtained). **e** An actual photograph of the rGO bioelectrode worn by a volunteer. SEM images of rGO bioelectrodes **f** before and **g** after 108 h continuously in contact with human skin after the in vivo study. (Reproduced from [36] with permission from ScienceDirect)

and muscles via electrical stimulation [55]. Analysing this data enables the identification of the functional status of muscles, peripheral nerves, neurons, and neuromuscular junctions. It is a crucial instrument in various fields, including biofeedback, ergonomics, motion analysis, neuromuscular function assessment, and diagnostic medicine [56]. Traditional human bioelectric signal sensing typically depends on disposable Ag/AgCl gel electrodes. However, the effectiveness of these conventional electrodes declines over time due to the drying out of the conductive gel, which raises the likelihood of skin inflammation [35, 57]. As a result, creating new flexible sensing electrodes has become a key focus in research. Graphene-based materials for developing electrodes for EMG have gained researchers' attention.

In 2019, Ozturk et al. investigated the ability of wearable graphene textiles for muscular activity monitoring and their applicability for surface electromyography (sEMG) [58]. The study comprehensively demonstrated the potential of graphene-based textiles as viable materials for wearable muscular monitoring devices. A dip-dry-reduce process was applied to prepare graphene-coated textiles. Modified Hummer's method was used to produce the GrO solution, and a piece of textile was dipped into the solution and then dried. A subsequent reduction was also applied to GrO to produce rGrO. The application of graphene textiles in wearable muscular monitoring demonstrates practicality, evidenced by the successful acquisition of sEMG signals, which were then compared to the performance of conventional wet silver/silver chloride (Ag/AgCl) electrodes. The evaluative comparison presented several critical parameters, including the SNR, cross-correlation, and the level of sensitivity to power-line interference. Despite the observation that graphene textile

Table 4.1 Features of Graphene-based non-invasive ECG sensing devices

Electrodes	Sheet Resistance (SR)/ Contact Impedance (CI)/ Conductivity ©	Flexibility	Other attributes
Gr/CNT/PDMS (dry) [38]	SR/CI:U[1]. C: 0.01 S.cm^{-1}	Stretchable and bendable	Adhesion strength: 1.3 N cm^{-2} adhesion strength
rGrO/PDMS (dry) [39]	SR/CI/C:U[1]	Attached to skin conformally	Adhesion strength:6.6 N cm^{-2} (dry surface), 5.3 N cm^{-2} (wet surface), 4.5 N cm^{-2} (under aquatic conditions), and water-proof
Gr/SEBS (dry) [40]	SR: 150 Ω.sq−1. CI/C: U[1]	More than 10 attaching and detaching cycles	83% transmittance 30 dB signal- to-noise ratio
CB/rGrO/SMA-M1000 (dry) [35]	SR: 2.5 × 101 Ω.sq^{-1}. CI: 4.6 kΩ (1:1:0), 0.19 kΩ (0:10:1), 0.38 kΩ (5:5:1), 4.6 kΩ (1:1:1), and 61 kΩ (1:1:5)	More than 3000 repeated uses	Mechanically durable (3000 cycles), Hydrophobic
Gr ink/ tex- tile electrode (dry) [41]	SR: 104 Ω.sq^{-1}. CI/C:U[1]	Flexible and twisted	130 g·m^{-2} (lightweight)
Gr/SF/Ca2 + (wet) [42]	SR: 0.62 to 0.28 MΩ.sq^{-1} over SR: 0.62 to 0.28 MΩ.sq^{-1} over 20 °C to 50 °C. CI/C:U1	Stretchable and compressible	100% healing efficiency in 0.3 s
PC/rGO/PVA hydrogel (wet) [43]	SR/CI/C:U[1]	>5000% stretchability	95.73% healing efficiency in 0.3 s
PVA/GNR Elec-trode (dry) [44]	CI: 250 kΩ for the 10–1 kHz frequency range. SR/C:U[1]	Flexible	59.99 dB signal–noise-ratio, water-soluble
CuNPs/PTT/GrO [45]	SR: 8.12 × 10^{-2} Ω sq^{-1} (initial), SR: 9.06 × 10^{-1} Ω sq^{-1} (after 30 days). CI/C:U[1]	Bendable	Highly conductive network, air- stable, washable
MoCl5 inter-calated bilayer graphene (Mo-BLG) [46]	SR: 40 Ω.sq^{-1} (initially), 60 Ω.sq^{-1} (after exposure to the atmosphere over 1 month). CI/C: U[1]	Sustained 80% strain	Durable over 2000 cycles at 30% strain
Gr/PHA (dry) [47]	SR: 20 k Ω. sq^{-1}. CI/ C:U[1]	Flexible	3-lead ECG sensor; electrodes could last for 1–2 years
laser-rGrO/PET (dry) [36]	CI: 4 kΩ. SR/C:U[1]	Flexible	70 dB signal-to-noise ratio; stability for up to 100 h in harsh conditions and upon skin contact

(continued)

Table 4.1 (continued)

Electrodes	Sheet Resistance (SR)/ Contact Impedance (CI)/ Conductivity ©	Flexibility	Other attributes
Gr/PVA (dry) [49]	SR: 75 Ω.sq^{-1}. CI: 45.12 kΩ at 100 Hz. C:U[1]	Young's modulus of 8.598 MPa	60 dB signal-to-noise ratio; may undergo delamination within 150 s upon deionised water application
PEP/3DPG (dry) [50]	CI: 198.2 kΩ at 4 Hz to 90.35 kΩ at 1 kHz. SR/ C:U[1]	Bendable	13.5 dB signal-to-noise ratio; areal capacitance of 16.94 mF·cm^{-2}; detection range 0.1–700 pg·mL^{-1} and LOD: 0.34 pg·mL^{-1}
LIG/MXene-Ti3C2Tx @EDOT [51]	CI: 51.08 kΩ at 10 Hz. SR/C:U[1]	High strain sensitivity (2075; > 22% strain)	20.14 dB initial signal-to-noise ratio and 19.04 dB after 8 h
GrO/GI/PVA (dry) [52]	SR/CI/C:U[1]	Plastically	85.21 mV (peak to peak) value for transparent electrodes, Long- term stability 30.22 μs skin conductivity

[1]U = Unknown

electrodes exhibited a higher susceptibility to power-line interference, they demonstrated similarities in performance with Ag/AgCl electrodes, particularly in terms of SNR and overall signal morphology. Notably, the correlation values for sEMG signals obtained from the biceps brachii muscle reached up to 97%.

Furthermore, in a 2021 study, Ozturk et al. captured sEMG signals to monitor localised muscle fatigue using a smart armband [59] made of graphene-coated textiles by applying the same procedure to synthesise the textiles as mentioned above. They achieved this using a custom-designed, small-scale front-end readout circuitry. The graphene-based textile demonstrated a correlation coefficient of approximately 97% compared to the benchmark Ag/AgCl electrodes. The SNR values for the Ag/AgCl and graphene textile electrodes were 15.9 dB and 14.3 dB, respectively. The experiment involved isometric contraction of the right biceps brachii of the subjects. The recorded signal indicated signs of localised muscle fatigue, displaying an apparent increase in total band energy and a decrease in instantaneous median frequency (IMF) of −0.19 Hz/s. The linear regression slope was measured as 0.0082 $mV^2\ s^{-1}$.

In their study, Song et al. presented a self-healing polymer (SHP) that was reported to be both stretchable and conductive, designed for use in an EMG sensor [53]. They achieved this by dropcasting a composite solution consisting of a rigid self-healing polymer (PDMS-MPU0.4-IU0.6, SHP) and graphene. The optimised composite initially exhibited a resistance of approximately 40.5 Ω and demonstrated resilience during a cyclic test of 200 stretches at 50% strain. It was found that for low graphene content in MPU0.4-IU0.6, SHP/Gr composite (1:03), the sheet resistance was higher (119.1 Ω sq^{-1}) and decreased with the increasing ratio of graphene, i.e. 40.5 Ω sq^{-1} for the ration 1:05, 38.0 Ω sq^{-1} for 1:0.7 and 37.1 Ω sq^{-1} for 1:09. However, it can be noted that the sheet resistance is similar among composite with ration (1:0.5), (1:0.7) and (1:0.9) indicating the conductivity could be saturated at a specific ratio of graphene. Due to the SHP matrix's low glass transition temperature, the graphene-SHP composite autonomously repaired itself during mechanical deformation of approximately 50% strain by reinstating electrical pathways in dynamic polymer networks. This impressive self-healing process occurred at room temperature, without the need for external sources. Moreover, the composite maintained stable electrical performance in cyclic stretching tests after self-healing without requiring a bilayer structure. Finally, the researchers demonstrated the monitoring of the EMG signal by applying an alginate hydrogel to coat the surface of the graphene-SHP composite, which was then utilised by a robot to replicate human hand motions.

In 2020, Das et al. proposed advanced paper-based epidermal sensors, employing biocompatible nylon membranes and non-toxic chemicals through simplistic methodologies [60]. They fabricated an epidermal sensor (TrGrO/NM) of thermally rGrO (TrGrO) integrated with a nylon membrane (NM). This sensor was developed to facilitate the application of conventional Ag/AgCl sensors. The fabricated sensor demonstrated a sheet resistance metric of 40 Ω sq^{-1}, coupled with a skin-contact impedance of approximately 20 kΩ at a lower frequency spectrum. Micro-gaps on the nylon membrane surface of the TrGr/NM sensor facilitated variations in resistance under both tensile and compressive strains. Furthermore, the sensor exhibited a response time of approximately 0.5 seconds. The practical applications of the TrGrO/NM sensor have been reported to include ECG, EEG, and EMG, as well as monitoring various human movements.

Li et al. researched the development of highly flexible and stretchable 3D-printed rGrO/polymer nanocomposites (Gr/SMA/PEA) for use in smart clothing for EMG and ECG applications in 2023 [54]. They created photocurable resins with enhanced stretchability and resilience by blending acrylate monomers and oligomers in various ratios. Graphene was added to improve conductivity and flexibility. The modified resin was used to 3D print sample patterns with microneedle surface structures, which served as sensors for capturing human body signals. To prevent graphene from clumping, two strategies were tested: the first incorporated styrene maleic anhydride (SMA) to facilitate physical adsorption through π–π interactions with graphene's aromatic rings, and the second added polyether amine (PEA) to create steric hindrance with graphene. An amphiphilic dispersant was synthesised by combining SMA with PEA, and the optimal dispersion ratio was determined. This dispersant was mixed with the photocured resin to produce a graphene/photocured resin nanocomposite. By adjusting the amount of graphene, they achieved a nanocomposite with low electrical resistance (around 3 10^3 Ω sq^{-1}) and robust mechanical properties (tensile strength: over 4 MPa, elongation at break: 320%). This material also demonstrated excellent durability under 50% cycle fatigue testing. Unlike traditional screen printing, they used 3D printing to create flat base plates. Various microneedle surface structures with different length to diameter ratios and needle spacings were fabricated, and their effectiveness in maintaining skin contact was evaluated by measuring electrocardiography and electromyography signals in human subjects. The findings suggested that customisable 3D-printed resins with intricate structures could be seamlessly integrated into smart garments for real-time monitoring of human physiological conditions.

Furthermore, in a 2024 study, Li et al. developed a flexible and stretchable photocurable resin for ECG and EMG smart clothing by adjusting the ratio of oligomer to acrylate monomer [61]. They enhanced the electrical conductivity of the resin by incorporating graphene and silver nanoparticles (AgNPs), resulting in lower resistivity than their previous research [54]. Through a photocurable 3D printing method, they developed electrode surface and line structures to detect physiological signals such as EMG and ECG. The researchers utilised PIB-SA and POE to create an amphiphilic polymer dispersant (PIB-POE-PIB). Graphene was dispersed through hydrogen bonding between PIB-POE- PIB and graphene, lone pair-π stacking, and hydrophobic effects as an oligomer containing urethane chains. They subsequently combined urethane acrylate (UAc) with photo-initiators and light-curing monomers to develop organic/inorganic hybrid photocurable resins. Finally, using DLP 3D printing, they fabricated electrodes with octopus-inspired structures and pre-stretched lines to improve skin contact area and stretchability. They optimised the oligomer-to-monomer ratio to create an elastic photocurable resin, achieving optimal mechanical strength at a 1:14 oligomer-to-monomer ratio [61]. By adding graphene and silver nitrate at a weight ratio of 1.5/15 wt%, the sheet resistance of the photocurable nanocomposite resin reached 1.2 $10^2\ \Omega\ sq^{-1}$, withstanding a strain of up to 300% at fracture. Even after washing and perspiration tests, the resistivity remained at 6 $10^2\ \Omega\ sq^{-1}$. Finally, they utilised photocurable 3D printers to produce electrodes with various surface and line structures for measuring the ECG and EMG of the human body during different exercises, demonstrating the successful integration of photocurable 3D printing with wearable electronic devices for intelligent textiles used in bioelectric signal detection [61].

In 2024, Pan et al. developed a stretchable and sweat-resistant epidermal electrode using graphene foam and a eutectoid mixture (PDESC) comprising poly(acrylic acid) (PAAc), a deep eutectic solvent (DES) and cetyltrimethylammonium bromide (CTAB) for EMG signal capture [62]. The CVD process was utilised to create graphene foam combined with eutectogel. The resulting copolymer exhibited stretchability of 500–600%, adherence to sweaty skin of 19 kPa, and impedance of 99 Ω. It served a dual purpose, providing both structural integrity and a conductive interface for the graphene foam, thereby reducing skin contact impedance.

4.2.3 Electroencephalography (EEG)

The EEG technique utilises advanced electronic tools placed on the scalp to measure the electrical activity of the brain [64]. This activity refers to the rhythmic and spontaneous activity of groups of brain cells, as recorded by electrodes [65]. Wearable e-textiles are gaining popularity for non-invasive health monitoring due to their potential. However, creating multifunctional wearable e-textiles remains challenging due to poor performance, comfort, scalability, and cost [65].

Golparvar et al. studied the potential use of rGrO/textiles for capturing neural biopotential signals from the forehead, comparing their effectiveness with that of traditional commercial dry electrodes [66]. In the study, rGrO-infused electronic textile sensors were used in initial experiments to record brain activity, including alpha rhythm, with a similarity of approximately 91% compared to standard dry electrodes. The electrical properties of the electronic textile, such as sheet resistivity (14 kΩ sq^{-1}), conductivity (0.33 S/m), and thickness (0.2 mm), were noted. Additionally, skin-electrode impedance levels were measured at 90 kΩ in DC, 59 kΩ at 50 Hz, and 35 kΩ at 100 Hz. After undergoing five handwashing cycles, the surface resistance of the fabric increased from 45 kΩ to 65 kΩ. These textile-based sensors were adaptable and lightweight, weighing approximately 50 mg.

In 2022, Islam et al. proposed a fully printed, highly conductive, flexible, and machine-washable e-textiles platform capable of storing energy and monitoring physiological conditions, including bio-signals [63]. The method involved the scalable printing of graphene-based inks (Gr ink) on a rough and flexible textile substrate (i.e. cotton), followed by encapsulation with a hydrogel-polymer electrolyte, PVA doped with H_2SO_4, to create a highly conductive, machine-washable e-textile platform. The resulting e-textiles (Gr ink/cotton/PVA/H_2SO_4 gel electrolyte) were reported to be flexible and conformal, capable of detecting the activities of various body parts. The printed in-plane supercapacitor presented an aerial capacitance of approximately 3.2 mF cm^{-2} with stability for about 10,000 cycles. They demonstrated that such e-textiles could record brain activity (EEG) and found them comparable to conventional rigid electrodes. This advancement could lead to a multifunctional garment made of graphene-based e-textiles that functions as flexible and wearable sensors, powered by the energy stored in graphene-based textile supercapacitors.

4.2.4 Electrooculography (EOG)

Electrooculography (EOG) non-invasively measures changes in the eye's electrical potential triggered by light adaptation [32]. It effectively monitors behavioural and physiological states, such as reading patterns, sleep disturbances (e.g., rapid eye movement irregularities), and drowsiness detection, by tracking eye movements via corneal-retinal potentials. Nevertheless, further innovation in the field is limited by the technical complexity and high production costs of electrodes, which hinder their wider manufacturing and scalability.

Golparvar et al. [68] developed e-textiles with a graphene coating using a similar dip-dry-reduce technique as reported in [66] and Ozturk et al. [58]. In testing the effectiveness of these textiles for EOG acquisition, they found that the e-textiles performed comparably to traditional Ag/AgCl electrodes. The sheet resistance was reported to be approximately 20 kΩ sq^{-1}. To assess the performance of the textiles, they were cut into 3 cm × 3 cm pieces and soldered to a backing with wires for electrical connection. The study involved sixteen EOG recordings from eight participants using graphene textiles and Ag/AgCl electrodes, yielding an average signal correlation of approximately 80% and a maximum of 87% for one participant. Interestingly, signals obtained simultaneously from two participants, each wearing a different electrode type, showed a correlation of 73%, demonstrating that graphene textile electrodes can capture EOG patterns associated with horizontal saccades, blinks, and fixations with a high level of similarity to Ag/AgCl electrodes despite differences in physiology, contact conditions, etc. The report suggested that the wearability, flexibility, stretchability, and cost-effective manufacturing of graphene e-textiles make them an exciting prospect for the future of EOG applications. Additionally, with further development, graphene textiles could be seamlessly integrated with regular clothing and electronics, allowing for battery-powered operation and revolutionising the field of wearable human-computer interfaces. Potential applications reported included monitoring epileptic patients and driver drowsiness, as well as diagnostic polysomnogram tests for sleep disorders.

Furthermore, in 2021, Golparvar et al. introduced a self-sustaining wearable technology featuring a headband with flexible graphene textile electrodes [69]. These electrodes represent an advancement over the traditional 'wet' electrodes by integrating compact, portable readout electronics that facilitate real-time signal analysis and enable data transmission via Bluetooth to a remote device. The fabrication of conductive graphene textiles was achieved through a cost-effective, scalable tripartite methodology that entails the deposition of conformal graphene layers onto various textile substrates, including nylon, polyester, cotton, and Kevlar. Initially, these fabrics were submerged in a solution of GrO, synthesised utilising a modified Hummers method, and subsequently dried to promote the adhesion of GrO layers onto the fabric's surface. The process proceeds with the application of reducing agents such as hydrazine or hydrogen iodide, followed by a final washing in distilled water, culminating in the creation of a durable graphene coating. These graphene-coated fabric segments, approximately 3 cm × 3 cm, were attached to flexible foam padding utilising metallic snap fasteners, thereby establishing connectivity with the device's electronic components. This approach forms 'passive' graphene textile electrodes primed for immediate deployment in acquiring bioelectrical signals without requiring further modifications. Conductivity assessments of these textiles have demonstrated resistances between 1 and 10 kΩ, with the impedance at the skin-electrode interface ranging from 87.5 kΩ at 10 Hz to 11.6 kΩ at 1 kHz.

Moreover, the ability of these textile electrodes was reported to increase in the presence of moisture and sweat, which served to enhance conductivity at the skin-electrode junction, thereby offering an improved SNR for extended periods of monitoring in comparison to traditional 'wet' electrodes, which exhibited a decrease in performance over time. Graphene textiles were proposed to be used in developing wearable eye-tracking technologies and in facilitating the control of remote objects through the interpretation of eye movements, including the manipulation of a mouse cursor for typing on a virtual keyboard and navigating a four-wheeled robot through a maze, accomplished using five distinct eye movements detected by a single-channel

EOG system. This configuration achieved typing speeds of up to six characters per minute without the use of prediction algorithms and has successfully guided the robot through a maze containing four 180° turns, achieving pattern recognition accuracies of 100% and 98%, respectively [69].

In a 2020 study, Li et al. introduced an approach to fabricating electrodes using rGrO, utilising a room-temperature reduction and patterning process [67]. The rGrO electrodes, featuring an adhesive backing, were reported to ensure a comfortable fit against the skin, maintaining low impedance at the skin-electrode interface, even in the presence of skin movement or deformation. The research team devised a dual-purpose method that enables simultaneous reduction and patterning of GrO without requiring high-temperature processing or intricate patterning techniques. This novel approach allows the scalable production of multi-electrode arrays through simplified fabrication processes, as demonstrated by the integration of soft lithography and flash-induced reduction. The electrodes produced using this method exhibited low impedance, highlighting their potential for use in high-performance applications. They maintained stability across high frequencies (100 to 1000 Hz), a critical range considering that EMG signals predominantly exhibited within this spectrum. Using modified Hummer's method, GrO was produced. An adhesive tape with electrode patterns was pasted on a silicon substrate. Chromium bonding and Copper layers were sputtered on the silicon substrate, and the mask was removed after the metal deposition. This silicon substrate was then immersed in GrO solution at 60 °C to obtain rGrO pattern. After further processing, the rGrO patterns were transferred onto PDMS. Optimal outcomes were observed with rGrO sheets derived from a concentration of 0.5 mg ml^{-1} of GrO over a two-hour reaction period, with impedance levels fluctuating between approximately 150 kΩ and 50 kΩ. The capability of these electrodes to capture eye movement signals across a spectrum of horizontal angles, from −60 to +60 degrees, alongside the notable variations in EOG amplitude at different angles, was reported [67]. Furthermore, a multi-channel array of rGrO electrodes was evaluated for its ability to discern five distinct hand gestures relevant to grasping actions. Utilising Fisher linear discriminant analysis as a binary classifier, the system exhibited a success rate exceeding 75% in controlling an advanced robotic hand. This electrode was reported to be capable of recording EOG, EEG, EMG, and ECG signals.

In their study, Beach et al. introduced an innovative EOG monitoring device that incorporated conductive and flexible graphene-based textiles into a commercially available eye mask [70]. The textiles, made from nylon, served as sensing electrodes and were created on a scoured and bleached nylon knitted fabric. The process involved reducing a GrO solution, its functionalisation, and coating the textile material. The modified Hummer method was employed to prepare GrO, which was then chemically reduced to create rGrO using sodium hydrosulphite ($Na_2S_2O_4$). The surface of rGrO is functionalised with PSS to ensure a stable dispersion when coating the textile material. The coated textile was dried at 100 °C for five minutes using an industry-standard pad-dry unit. The resulting textile material was reported to exhibit a sheet resistance below 40 kΩ sq^{-1}. The mask was tested on four participants to quantify the noise floor and demonstrate its ability to detect eye blinks at a high SNR of over

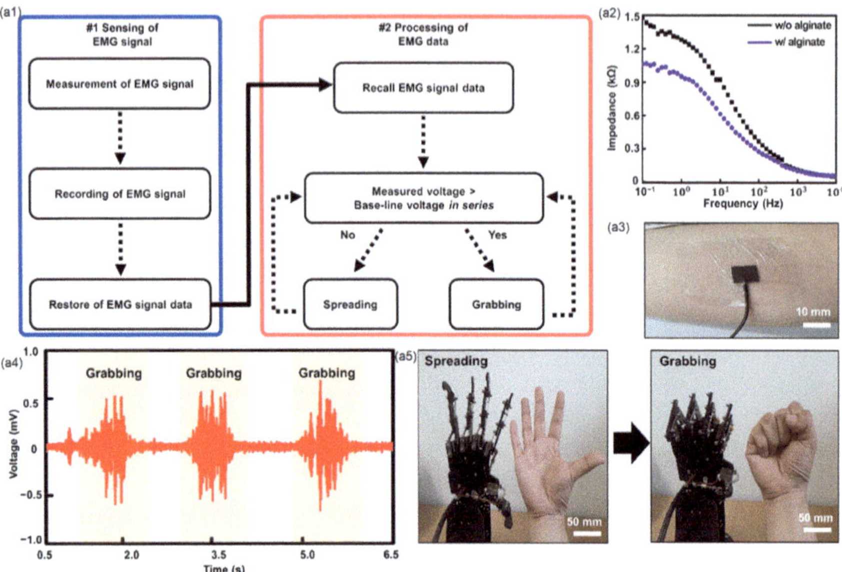

Fig. 4.3 Demonstration of human–robot interface with stretchable and self-healing electrodes. **a1** Flow diagram of EMG signal recording and processing. **a2** Impedance data of the composite with and without alginate. **a3** A photograph of the electrode attached to the skin. **a4** Recorded EMG signals using the electrode. **a5** Demonstration of the humanrobot interface (spreading and grabbing). (Reproduced from [53] with permission from MDPI)

16 dB. Additionally, it was found that the material can detect other eye movements with low noise. The system was held in place with only an elastic strap, providing the same level of comfort as wearing a regular eye mask. They may be beneficial for comfortable eye movement experiments, particularly during sleep studies, where bulky instrumentation is routinely used to monitor EOG. Figures 4.3, 4.4, 4.5, 4.6, and Table 4.2 present different electrophysiological sensors (e.g., EMG, EEG, and EOG) along with their features.

4.3 Kinematic

There have been many attempts to develop flexible and stretchable sensors that can detect and monitor human motion. These sensors can capture essential signals, including movement and respiratory disorders, blood pressure, and athletic performance tracking data. Sensitive pressure, tactile, or strain sensors can be integrated to monitor various body parts continuously. Multiple structures and sensing materials are utilised to create flexible, sensitive pressure, tactile, and strain sensors, enabling these achievements. Depending on the functional material used, various sensing mechanisms are required for skin-integrated flexible sensors, such as piezoresistive,

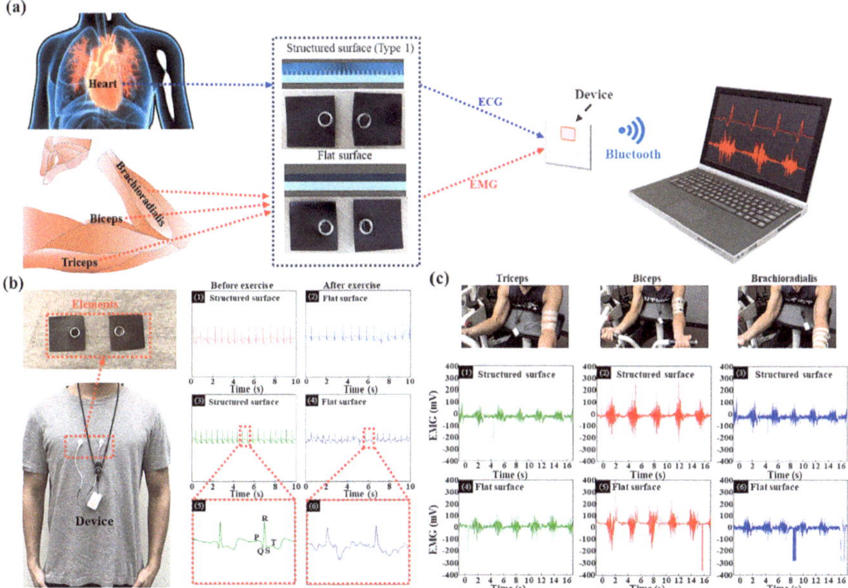

Fig. 4.4 **a** Sensing electrodes were attached to the skin near the heart and upper arm, and signals were transmitted to a laptop computer via a Bluetooth device; **b** schematic diagram of the use of electrodes with a structured surface (1, 3, 5) and flat surface (2, 4, 6) for ECG measurements before exercise (1,2) and after exercise (3, 4), with magnified signals shown in (5) and (6); **c** attachment of sensing electrodes with a structured surface (1,2,3) and flat surface (4, 5, 6) to the triceps (1, 4), biceps (2, 5), and brachioradialis (3, 6) for EMG signal measurement. (Reproduced from [54] with permission from ScienceDirect)

piezocapacitive, piezoelectric, piezophotronic, and triboelectric types. Among these mechanisms, piezoresistive and piezocapacitive designs, which measure changes in resistance and capacitance, are commonly used due to their simple design and direct data acquisition capabilities. Graphene has emerged as a promising material for pressure, tactile, and strain sensors, highlighting recent advances in this field [17].

4.3.1 *Acoustic Sensor*

Graphene-based acoustic sensors utilise mechanical vibrations to detect sound [71, 72]. The electron mobility of graphene allows it to convert acoustic signals into electrical ones through the piezoresistive effect, enhancing sound detection. Graphene also plays a role as a sound emitter and utilises its thermal conductivity to convert electrical signals back into sound through the thermoacoustic effect. Both single-layer and multi-layer graphene are superior in acoustic applications, providing a broader frequency response and minimising variations compared to traditional sound devices.

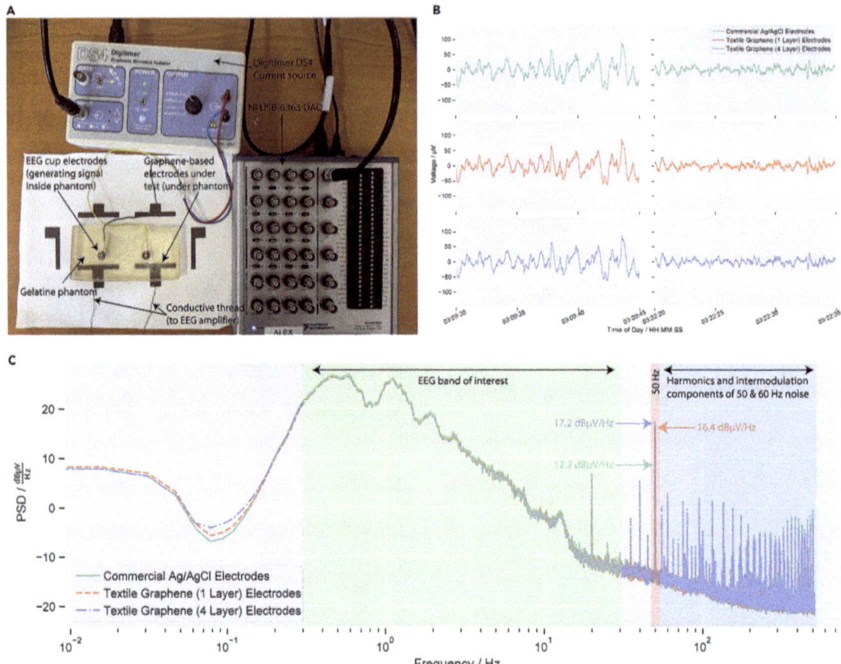

Fig. 4.5 Printed graphene e-textiles for EEG applications (**A**) Experimental setup showing signal source (DAQ), current stimulator (Digitimer DS4), gelatine phantom, graphene-based electrodes, and conductive thread connection to EEG Amplifier. **B** Data collected using the textile graphene-based electrodes against commercial Ag/AgCl electrodes from the Fpz-Oz data in the SC4001 record from Physionet. (left) A section of the record is in the deepest stage of sleep (stage 4), indicated by the presence of slow oscillations in the record. (right) A lighter stage of sleep (stage 2 is indicated by smaller amplitude, higher-frequency oscillations). **C** Power spectral density of the recorded signal from Graphene Textiles electrodes compared to commercial Ag/AgCl electrodes, highlighting the typical frequency band of interest in EEG studies, shaded in green. The 50 Hz component and its contribution for each electrode type are shaded in red. Also shown are higher-frequency harmonics and intermodulation components from the 5–0 to 60-Hz noise, shaded in dark blue. (Reproduce with [63] permission from ScienceDirect)

Graphene-based thermoacoustic (TA) loudspeakers, which utilise a graphene film attached to a substrate to generate sound through heat, are contributing to the enhancement of their thermoacoustic efficiency. In 2017, Xing et al. refined the existing TA model to account for the influence of small thermal wavelengths and developed an acoustic platform for validating this model [73]. Furthermore, the team investigated the impact of the thickness of the graphene film layers, the substrate's thermal properties, and the types of gases employed on thermoacoustic efficiency. The results showed that the thickness of the graphene films, ranging from 5 nm to 15 nm, had a negligible effect on efficiency and presented a sensitivity coefficient of less than 0.02. In contrast, the thermal responsiveness of the substrate emerged as a significant

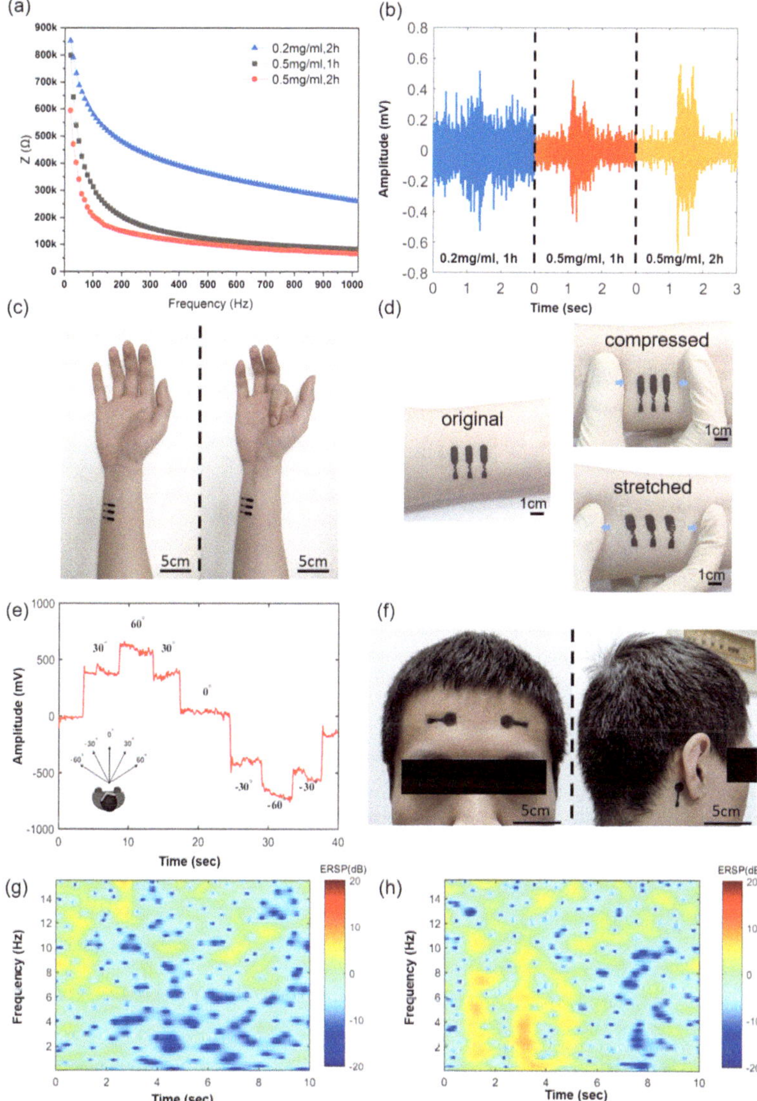

Fig. 4.6 Measurement performances of the rGO electrodes. **a** The contact impedance between different electrodes and human skin. **b** EMG acquired by rGO electrodes without electrode gel. **c** Images showing the locations of the electrodes and the corresponding hand motion during EMG measurements in (**b**). **d** The conformal contact at the skin-rGO electrode interface during compression and stretch. **e** EOG was acquired from the side of the eye with different gazing angles and the locations of rGO electrodes. 'A' and 'B' are two differential electrodes, and 'G' is the ground electrode (Fig. S7). **f** Image of differential electrodes (left) and ground electrode (right) mounted on the subject. **g, h** EEG acquired from the forehea$_1$d$_4$ during repose (**g**) and concentration (**h**). (Reproduced from [67] with permission from ScienceDirect)

Table 4.2 Features of Graphene-based non-invasive EMG, EEG, and EOG sensing devices

Sensor: Electrode type	Sheet resistance (SR)/ Contact impedance (CI)/ Conductivity (C)	Flexibility	Other attributes
EMG: rGrO/textile [58]	SR/CI/C:U[1]	Flexible and Stretchable	Signal–noise-ratio – 26.7 dB, corre- lation coefficient against Ag/ AgCl electrodes is ~97% (~0.88)
EMG: rGrO/textile [59]	SR/CI/C:U[1]	Bendable	Linear Regression line slopes: 0.0082 mV2 s – 1. SNR is 14.3 dB, correlation coefficient against Ag/AgCl electrodes is *sim* 97%
EMG: PDMS-MPU0.4-IU0.6, SHP/Gr[53]	CI/C:U[1]. SR: 40.5 Ω sq^{-1} for $MPU_{0.4}$-$IU_{0.6}$. SHP/Gr ratio of 1:0.5	Stretchable	Self-healing during mechanical deformation
EMG: TrGrO/NM [60]	SR: 40 Ω sq^{-1}. CI: 20 kΩ. C:U[1]	Flexible, bio-compatible, and reusable	0.5 s response time
EMG: rGrO/SMA/ PEA [54]	SR: 3×10^3 $Ω.sq^{-1}$. CI/C:U[1]	Flexible and stretchable	Tensile strength: over 4 MPa, elongation at break: 320%, dura- bility under 50% cycle fatigue testing
EMG: AgNPs/Gr/ PIB- POE-PIB [61]	SR: 6×10^2 $Ω.sq^{-1}$. CI/C:U[1]	Flexible and stretchable	Strain of up to 300% at fracture
EMG: Gr/PDESC [62]	CI: 99Ω. SR/C: U[1]	Stretchable	Stretchability (500%–600%), adherence to sweaty skin (19 kPa)
EEG:rGrO/textile [66]	SR: 14 $kΩ.sq^{-1}$. CI: 90 kΩ (DC), 59 kΩ (at 50 Hz), and 35 kΩ (at 100 Hz). C: 0.33 S/m	Flexible	Textile thickness of ~0.2 mm: flexible, weighing ~50 mg
EEG:Gr ink/cotton/ PVA/H2SO4 gel electrolyte[63]	SR/CI/C:U[1]	U[1]	Stability ~10,000 cycles, flexible, washable resistance 30Ω/cm, Areal capacitance 3.2 $mF \cdot cm^{-2}$
EOG:rGrO/textile [68]	SR: 20 $kΩ.sq^{-1}$. CI/ C:U[1]	Gel-free, flexible, dry textile electrodes	Can capture EOG patterns asso- ciated with horizontal saccades, blinks, and fixations

(continued)

Table 4.2 (continued)

Sensor: Electrode type	Sheet resistance (SR)/ Contact impedance (CI)/ Conductivity (C)	Flexibility	Other attributes
EOG:rGO/textile [69]	CI: 87.5 kΩ at 10 Hz to 11.6 kΩ at 1 kHz. SR/ C:U1	Flexible, seamless	Pattern detection accuracies of 100% (without prediction algo- rithms) and 98% (with guidance of the robot)
EOG:rGrO/PDMS [67]	CI: 200 kΩ at 100 Hz to 50 kΩ at 1000 kHz. SR/C:U1	Flexible	High-quality signals even without electrode gel
EOG:rGO/textile [70]	SR: below 40 kΩ. sq^{-1}. CI/C: U[1]	Flexible and conductive	SNR: 16.5 dB

[1]U = Unknown

factor, exhibiting a notable sensitivity coefficient of 1.7. Substrates characterised by lower thermal responsiveness demonstrated improved acoustic performance [73].

Regarding the influence of ambient gases, the higher values of the sensitivity coefficients for density and thermal conductivity contribute to improved efficiency, and the sound pressure level produced by the graphene films exhibits an inverse relationship with specific heat [74]. These insights play a pivotal role in the design of more efficient graphene thermoacoustic speakers.

Moreover, graphene is increasingly recognised for its potential in thermoacoustic (TA) applications due to its high thermal conductivity and low heat capacity per unit area. However, its broader use in TA devices remains limited by constraints related to device size and the high costs associated with its manufacture. In 2024, Hou et al. proposed an economical dip-coating technique for creating a large-scale graphene-based auditory device [75]. The process involved applying graphene nanoplates to a polyurethane (PU) base, resulting in a highly flexible graphene foam (GrF) device. This GrF device was reported to produce up to 70 dB sound levels from a 1 mm distance with just 1 W of input power. Additionally, a thermoacoustic analysis model was developed to simulate the acoustic properties of the GrF, taking into account its unique porous design. The Grf was manufactured in sizes of 1.5 cm × 1 cm with a thickness of 5 mm. It was reported that the interaction between graphene and air was enhanced, and heat dissipation improved because of the three-dimensional porous structure of the device. The sensor's adaptability was reported to enable its use in fully flexible in-ear headphones, demonstrating the diverse potential applications for graphene-based loudspeakers.

Conventional sound production and detection devices are usually separate components that operate within the audible range for humans. The demand for smaller devices that can be integrated into wearable technology has led to the necessity for a single device capable of both producing and detecting sound. In 2017, Tao et al. introduced a concept known as the LIG artificial throat, which can perform both

Table 4.3 Features of Graphene-based acoustic sensors

Sensing material	Driven energy	Thickness	Dimension (cm × cm)	Sound pressure (dB)	Measuring distance (cm)	Frequency range (kHz)
Graphene/ PET, PDMS, Glass[73]	Electricity	5–15 nm	2 × 2	U[1]	1	U[1]–20
GrF [75]	Electricity	5 mm	1.5 × 1	52	3	7.5–15
Gr/PI [76]	Electricity	8–60 μm	1 × 2	90	3	0.1–40

[1]U = Unknown

functions [76]. This device was created through the direct laser patterning of polyimide (PI), departing from traditional acoustic transducers that rely on piezoelectric effects. The LIG artificial throat can emit a broad spectrum of sounds, ranging from 100 Hz to 40 kHz. Thinner LIG layers were observed to result in higher sound pressure levels. As a detector, it exhibited reactions to various sound types and throat vibrations, allowing it to differentiate between coughs, hums, and screams across a range of tones and volumes and even to recognise words and sentences. Due to its dual functionality, the LIG artificial throat can assist individuals with speech impairments by converting varying throat vibrations into specific, pre-designed sounds. Additionally, the production process of the LIG artificial throat was characterised by simplicity, high efficiency, outstanding flexibility, and low cost.

Table 4.3 summarises the features of graphene-based acoustic sensors.

4.3.2 *Pressure*

Yang et al. (2018) opined that graphene and its derivatives are popular materials for pressure sensors due to their impressive piezoresistive performance and ease of handling [77]. Lv et al. also noted in 2017 that flexible graphene-based pressure sensors with various composite materials and structures have achieved exceptional sensing properties [78]. These sensors can be used to monitor heartbeats, wrist pulses, radial arteries, respiration, phonation/acoustic waves, walking states, finger bending, wrist blood pressure, and facial expressions. Applying pressure sensors that can be efficiently adjusted to conform to varying curved surfaces has garnered considerable interest. These sensors help emerge fields such as wearable electronic skins and human-machine interfaces. Among the various types of pressure sensors, piezoresistive pressure sensors based on graphene have garnered considerable attention due to their exceptional electrical conductivity and nanoscale flexibility [79]. However, conventional piezoresistive pressure sensors tend to exhibit low sensitivity and thus may be unable to detect low-pressure conditions. To overcome this limitation, researchers have been exploring an alternative approach that involves using micro-structured, micro-patterned, and porous-structured sensing materials. This approach has demonstrated the potential to improve the sensitivity and accuracy of pressure sensors and is currently the focus of significant interest across both academic and industrial research. Researchers have explored techniques based on various electrical conduction mechanisms to enhance sensor performance [80–86]. However, their complex fabrication and high production costs currently limit their applications. These methods include

- A millefeuille-like architecture,
- Lotus-leaf-inspired hierarchical structures,
- Graphene elastomers,
- Nanowire/graphene heterostructures,
- Spinosum microstructures,
- Foams, and
- Sparkling blocks.

In 2018, Huang et al. proposed a piezoresistive pressure sensor featuring a distinctive millefeuille-like design [84]. This design integrated rGrO layers separated by covalently bonded molecular pillars of triethylene glycol amine ($C_6H_{14}O_4$), 1-octylamine ($C_8H_{19}N$), and 4-aminobiphenyl ($C_{12}H_{11}N$), offering customisable mechanical properties. Under minimal pressure, this layered structure experi- enced a decrease in electrical resistance due to electron tunnelling across the rGrO sheets, and the stiffness of the molecular pillars enabled adjustments to the sensor's sensitivity. This device exhibited sensitivity levels of up to 0.82 kPa^{-1} within a low-pressure range of 0–0.6 kPa, paired with rapid response times of approximately 24 ms and a low detection limit of 7 Pa.

Similarly, Pang et al. developed a mechanical sensor incorporating a graphene porous network (GPN) within PDMS in 2016 [81]. The sensor was created using a nickel foam template via a chemical etching process, enabling the integration of the GPN into the graphene-coated PDMS-nickel foam. It was reported that, due to its unique structure, the device could effectively detect pressure and strain, thereby enhancing the conductivity of the porous design. The sensor exhibited a sensitivity of 0.09 kPa^{-1}, capable of detecting pressure across a broad range of up to 1000 kPa. In addition, it demonstrated versatility in movement analysis, including identifying walking patterns, measuring finger bending, and monitoring wrist blood pressure.

Furthermore, in 2018, Pang et al. developed an approach to creating a surface design characterised by a random distribution of spinosum microstructure of PDMS integrated with rGrO [80]. Due to this design, the graphene-based pressure sensor was reported to achieve a sensitivity, of 25.1 kPa^{-1}, while ensuring linear performance within the 0–2.6 kPa pressure range. The comparative analysis indicated that this pressure sensor outperformed earlier designs featuring surface modifications [81]. Simulations and mechanistic analyses demonstrated that the random distribution and inoculum microstructure were crucial in enhancing the sensor's sensitivity and expanding its linear response range. Furthermore, the sensor enables real-time monitoring of heartbeat, respiratory functions, phonation, and various physical activities [80].

Chen et al. (2017) introduced a pressure sensor made from wrinkled graphene (WG) [87]. This material was created using a simple liquid-phase shrink method. The sensor (WG/AAO/WG) on PDMS consists of two layers of wrinkled graphene positioned opposite each other and separated by a thin porous anodic aluminium oxide (AAO) membrane, which acts as an insulator. The sensor demonstrates a sensitivity of 6.92 kPa^{-1} at lower pressure ranges (300 Pa to 1.5 kPa) and 0.14 kPa^{-1} at higher pressure ranges (1.5 to 4.5 kPa).

Moreover, in 2017, Chen Z. et al. proposed a pressure sensor that utilised a layered structure combining $PbTiO_3$ nanowires and graphene and encapsulated by PDMS to measure static pressures [82]. This sensor utilised the combined effects of strain-induced polarisation charges in the piezoelectric nanowires and the resultant changes in carrier scattering within the graphene. The sensor exhibited a sensitivity of up to 9.4 10^{-3} kPa^{-1} and a response time ranging from 5 to 7 μs.

Lou et al. developed a self-assembling 3D film platform that combines a naturally viscoelastic material, P(VDF-TrFe), with rGrO [88]. By layering these composite films between two unmodified PDMS films, they created a pressure sensor that demonstrated sensitivity at 15.6 kPa^{-1}, a response time of 5 milliseconds, and stability across 100,000 cycles. The sensor exhibited a detection limit of 1.2 Pa while operating on 1.0 V, making it sensitive enough to detect the touch of a feather or a grain of rice.

In 2017, another research group developed a hollow structure of 3D graphene/PDMS by tuning the material's electrical conductivity and mechanical elasticity by adjusting the number of graphene layers and modifying the composition ratio of PDMS [89]. This approach enhanced the sensor's sensitivity to 15.9 kPa^{-1} within a linear range of 60 kPa. Additionally, they improved the response time to 1.2 ms, including the rising time. By fine-tuning the graphene network and the thickness of the composite, they achieved stable and accurate sensor output with a measurement error of less than 6%, even when subjected to bending across a range of −25 mm to +25 mm bending radius.

Meanwhile, Zhang et al. presented a method for developing piezoresistive pressure sensors using multilayer structures, employing a one-pot synthesis technique in 2018 [90]. This method required a composite material, GSH@PET, which was directly formed by combining thiolated graphene (GSH) and polyester fabric (PET). The sensor, benefiting from the fabric's rough layers and the superior conductivity of GSH, demonstrated the capability to measure a wide pressure range from 0 to 200 kPa. Even after undergoing 500 loading and unloading cycles, it exhibited sensitivities of 8.37 kPa^{-1} and 0.028 kPa^{-1} for pressure ranges of 0–8 kPa and 30–200 kPa, respectively.

In 2018, Tewari et al. proposed a straightforward and economical approach to developing a pressure sensor by utilising hydrophilic rGrO ink combined with MWNTs [91]. This method involved coating polyurethane (PU) foam with the MWNT-rGrO ink. The resulting MWNT-rGrO@PU foam-based sensors can detect small-scale and large-scale movements, making them exceptionally versatile. When subjected to low pressure (below 2.7 kPa, 50% strain), the formation of microcracks that disrupt the flow of electrical charges results in a significant increase in resistance, allowing for the detection of small-scale motion. Conversely, at higher pressure, the compressive contact of the coated faces of the PU foam resulted in an abrupt decrease in resistance, which could be suitable for monitoring large-scale motion. The sensor also demonstrated flexibility and reproducibility over 5000 cycles. The reported sensitivity of the sensor is 0.022 kPa^{-1} (0–2.7 kPa), 0.088 kPa^{-1} (2.7–10 kPa), and KPa^{-1} (Above 10 kPa).

Zhang et al. [92] reported a piezoresistive sensor with high precision, fast response time, and excellent repeatability. The sensor was created by attaching graphene nanomaterials to a sponge substrate made of melamine, with a flexible eco-flex material serving as the substrate. This sensor can accurately measure a wide range of pressures, with high accuracy within a smaller range. To create the sensor, a mixture of 0.2 mg of graphene and 20 ml of absolute ethanol was processed under an ultrasonic dispersion operation for 40 minutes to form a graphene dispersion and then used to soak a melamine sponge (MS), which was dried in an oven and repeated until the entire sponge block was black. The sensing material was prepared, and polylactic acid (PLA) was used to create 3D-printed strain sensor moulds. The moulds were then solidified in the sponge with graphene ethanol dispersion. The resistance values of the graphene sheets changed as the gap between the sponge skeletons decreased. As the pressure increased, the sensor resistance decreased, returning to its original shape as the pressure decreased. The sensitivity of the sensor was reported to be 0.186 kPa^{-1} within the pressure range of 0 to 5 kPa. In testing, the results were compared to those from the Omron sphygmomanometer, with an error rate between blood pressure readings of 60/100–90/135 mmHg not exceeding 3%.

Peng et al. designed a pressure sensor using porous graphene (PG) to monitor cardiovascu lar status without interruption [93]. Ink printing technology created the sensor, making it suitable for large-scale production. The sensor (PG/PET/EVA) was protected by a polyethylene terephthalate/polyethylene vinyl acetate (PET/EVA) laminated film to enhance its durability, shielding it against unexpected shear forces on the skin surface. The sensor exhibited a sensitivity of 53.99 MPa^{-1}, a resolution of less than 0.3 kPa, and was capable of detecting pressures ranging from 0.3 kPa to 1 MPa. It was also reported to be robust and repeatable for up to 1000 cycles.

Yue et al. developed a pressure sensor made from rGrO film [94]. To develop rGrO from GrO, Vitamin C was used. The rGrO films were then encapsulated by the PET film. This sensor (rGrO/PET) was reported to be compact, lightweight, and flexible. The structure of the graphene film included surface irregularities and a porous, layered appearance in cross-section. They achieved this structure using an in situ chemical reduction technique that employed Vitamin C as a reducing agent. The pressure sensor was reported to measure 10.39 kPa^{-1} for pressures below 2 kPa and operate over a range of up to 200 kPa with a sensitivity of 0.0034 kPa^{-1}.

4.3.3 Motion Detection Sensor

Accurate human motion monitoring requires sensors that offer a wide measurement range, flexibility, durability, and sensitivity to sound [95, 96]. The graphene textile sensor can detect various movements, including finger, wrist, knee, and elbow bending.

Qiao and the team have created a state-of-the-art multilayer graphene epidermal electronic skin through laser scribing graphene [97]. Attaching the sensor to the mask and throat can effectively monitor respiratory signals and differentiate between strong, slow, weak or fast breathing. Additionally, research indicates that graphene-paper pressure can sense even the slightest changes in pressure, such as pulse and gas pressure [98]. The sensor can identify changes in breathing before and after exercise.

The utilisation of cost-effective commercial polyester non-woven fabrics (PNWFs) of varying sizes as substrates for sensor applications was investigated by Lu et al [99]. To fabricate a piezoresistive pressure sensor, the team employed a coating of rGrO and polydimethylsiloxane (PDMS) on the PNWF substrate. A subsequent heat treatment was applied to increase the conductivity, achieving an impressive conductivity rate of 0.29 $S.m^{-1}$. The resultant rGrO/PDMS-PNWF pressure sensors demonstrated performance in monitoring various physiological movements, including bending a single finger and movements of the arm and knee. These sensors can distinguish between gait speeds and jogging, and even detect blood flow pulsation when worn on the limbs. The sensors demonstrated a sensitivity of 23.41 kPa^{-1} within the 0 to 3 kPa range, dropping to 0.937 kPa^{-1} in the 3 to 30 kPa pressure range. Furthermore, the sensors exhibited a response time of 120 ms and a recovery time of 240 ms, with detection resolutions of 18 Pa for compression and 21 Pa for bending modes, accompanied by minimal hysteresis of 0.8%.

Researchers engineered a sponge by combining rGrO with polyaniline nanowires (PANI NWs) [100]. This process involved coating rGrO and synthesising PANI NWs on the backbone of the sponge to develop rGrO/polyaniline-wrapped sponge (PANI WS). This technique for creating pressure sensors (rGrO/PANI WS), which was reported to monitor human movements, incorporated the advantage of the Coulomb force between the positively charged PANI NWs, developed directly on the negatively charged rGrO. This approach was reported to enhance the microstructure of the 3D graphene sponge, resulting in a hybrid sponge that is lightweight and highly flexible. The sensor demonstrated adjustable sensitivity levels ranging from 0.042 to 0.152 kPa^{-1}, an operating range of 0 to 27 kPa, a response time of 96 milliseconds, and a lifespan of over 9,000 cycles. Additionally, it was reported that, with the rGrO/PANI WS, the sensor could detect subtle physiological actions such as voice recognition, swallowing, various mouth movements, and significant human motions like finger bending, elbow movement, and knee exercises in real-time. Figures 4.7, 4.8, 4.9 present different types of graphene-based non-invasive kinematic sensors for sensing blood pressure, pulse, motion, finger tapping, and elbow, wrist and knee bending.

Das et al. introduced the development of a self-powered triboelectric nanogenerator (TENG)- based pressure sensor by combining laser-ablated graphene (LAG) and

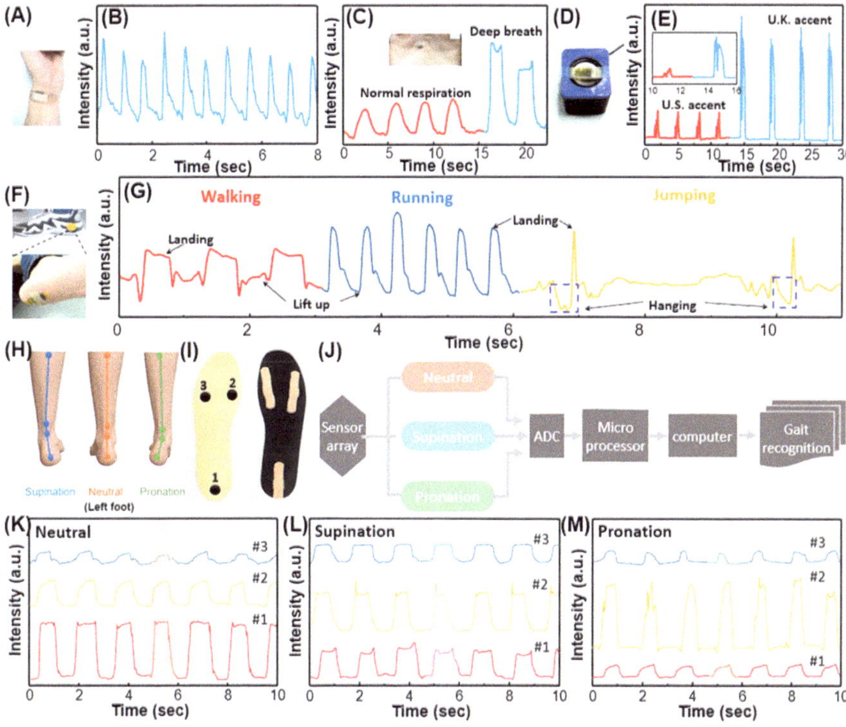

Fig. 4.7 Applications of the RDS graphene pressure sensor for various physiological signals' detection. Photograph of the pressure sensor assembled on (**A**) a wrist and corresponding signals of the (**B**) wrist pulse. **C** Photograph of a pressure sensor attached to the human chest area, indicating different signal variations for normal and deep respiration. Photograph of the pressure sensor attached to a (**D**) loudspeaker and corresponding signals (**E**) when the word 'graphene' was phonated, showing the ability to distinguish U.S. and U.K. pronunciation. **F** Photograph of the sensor put on the heel of the foot and (**G**) its detected signal when walking, running, and jumping. **H** Illustration of foot states for supination, neutral, and pronation and (**I**) photograph of pressure sensors fixed on the insole. **J** Schematic diagram of the system design for gait detection. Signals of the pressure array to detect gait states of **K** neutral, **L** supination, and **M** pronation, respectively. (Reproduced from [80] with permission from ACS Publications)

abrasive sandpaper [101]. This sensor, known as LAG-TEPS, consisted of a PDMS layer with LAG attached to the bottom and a PET/ITO film. The sensor demonstrated sensitivities in different pressure ranges, with a response time of 9.9 ms. It exhibited high flexibility and exceptional durability, enduring over 4,000 cycles of contact-separation tests. The researchers conducted evaluations of the LAG-TEPS in various areas of the human body, including the elbow, knee, and wrist, to assess its ability to detect human gestures. Through extensive bending and stretching movements, the sensor generated self-powered signals that capture diverse body motions and convert them into electrical signals, eliminating the need for an external energy source. Additionally, the LAG-TEPS was reported to distinguish between human

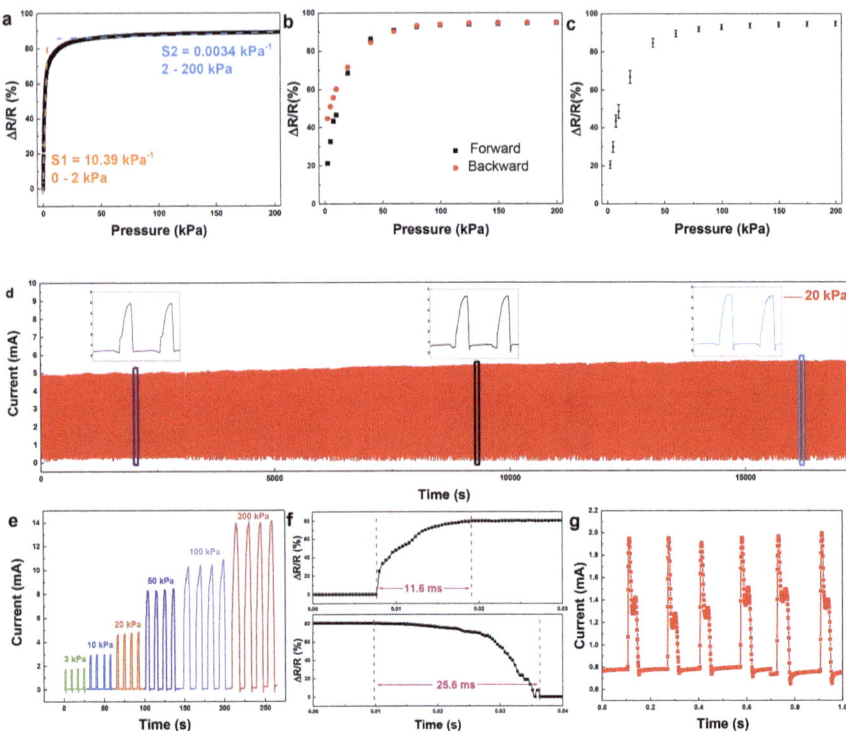

Fig. 4.8 **a** Resistance changing rate of GFPS response to pressure. **b** The hysteresis of GFPS. **c** The sample-to-sample error plot of GFPS. **d** The current of GFPS corresponding to 1100 cycles of repeated pressure, and enlarged plots corresponding to 2 cycles at 2000s, 9000 s, and 17,150 s, respectively. **e** The current response of GFPS to different repeated pressures. **f** Graphical plots of response (up) and recovery (down) time of GFPS. **g** Plot of GFPS response to a 6 Hz 5 kPa pressure loading–unloading cycle. (Reproduced from [94] with permission from ScienceDirect)

body gestures and detect pulses from human fingertips without external power. Table 4.4 summarises the features of graphene-based non-invasive kinematic sensing devices.

4.4 Body Temperature Sensors

Graphene-based temperature sensors measure the temperature difference between a person's body and their environment. The temperature data can indicate an individual's physical well-being [102–104]. These sensors are typically made from pristine graphene or rGrO because of their high thermal conductivity and emissivity [105,

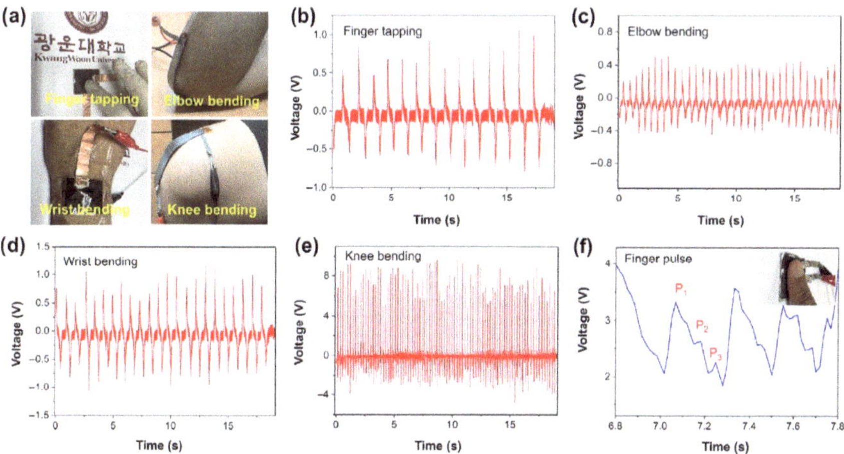

Fig. 4.9 Output voltage responses to repeated contact/separation of the flexible TEPS: **a** photograph of the LAG-TEPS placed on different parts of the body for the detection of human gestures. The corresponding voltage response for **b** finger tapping, **c** elbow bending, **d** wrist bending, **e** knee bending, and **f** finger pulses. (Reproduced from [101] with permission from Springer Nature)

106]. Various wearable temperature sensors can be produced using graphene structures, such as graphene fibre [107], 2D graphene networks [108–110] and graphene nanocomposites, which provide excellent stretchability and flexibility.

Hou et al. developed a dual sensor capable of measuring strain and temperature. The sensor utilised a nanocomposite hydrogel called SNGP, which is composed of sodium alginate (SA) nanofibrils, GrO, and polyacrylamide (PAM) [111]. Additionally, the inclusion of GrO enhanced the mechanical properties and the thermal responsiveness of the hydrogel. The fabrication process of the SNGP nanocomposite hydrogel involves blending a pre-treated SA/NaCl mixture with GrO, a monomer, a cross-linker, an accelerator, and a thermoset initiator. This mixture then underwent a 3-hour curing phase at 50 °C. The resulting hydrogel was encapsulated in a polyethylene terephthalate (PET) film or a high-bond adhesive tape to create the sensor. This sensor exhibited a decrease in electrical resistance with increasing temperature and a sensitivity of 2% °C^{-1} across a temperature range of 25–65 °C, surpassing the sensitivity of conventional ionic hydrogel-based sensors. The heightened sensitivity was likely attributed to electron movement between adjacent graphene layers [42]. The temperature sensitivity increased from 2.0% °C^{-1} to 2.7% °C^{-1} with a strain enhancement from 0% to 200%. The sensor also exhibited mechanical properties, including a tensile strength of 0.54 MPa, an elongation capacity of 3370%, and a compression strength of 4.4 MPa. The study revealed that a higher GrO concentration increased the sensor's toughness. The strain sensor achieved a gauge factor of 4.2 under a tensile strain of 2000%, capable of detecting minimal strains of 0.02% at a modest voltage of 0.5 V. The superior electrical characteristics of GrO were crucial in achieving this increased sensitivity.

Table 4.4 Features of Graphene-based non-invasive kinematic sensing devices

Sensing material	Type	Sensing mechanism	Sensitivity	Detection range (kPa)
rGrO/PDMS [80]	Motion/Pressure	Piezoresistive	25.1 kPa^{-1}	0–2.6
GPN/PDMS [81]	Pressure	Piezoresistive	0.09 kPa^{-1}	0– ~1000
rGrO/$C_6H_{14}O_4$	Pressure	Piezoresistive	0.82 kPa^{-1}	0–0.6
/$C_8H_{19}N$/ $C_{12}H_{11}N$ [84]				
WG/AAO	Pressure	Piezoresistive	6.92 kPa^{-1}	0.3–1.5
/WG/PDMS [87]			0.14 kPa^{-1}	1.5–4.5
$PbTiO_3$/Gr/ PDMS	Pressure	Piezoelectric	9.4×10^{-3} kPa^{-1}	0–1.4
rGrO/ P(VDF-TrFe)	Pressure	Piezoresistive	15.6 kPa^{-1}	20–55
3D Gr/PDMS [89]	Pressure	Piezoresistive	15.9 kPa^{-1}	0–60
GSH@PET [90]	Pressure	Piezoresistive	8.37 kPa^{-1} (at	0–200
			0–8 kPa)	
			0.028 kPa^{-1} (at	
			30–200 kPa)	
MWNT–	Pressure	Piezoresistive	0.022 kPa^{-1} (at	0 – > 10
rGrO@PU [91]			0–2.7 kPa)	
			0.088 kPa^{-1} (at	
			2.7–10 kPa)	
			0.034 kPa^{-1}	
			(above 10 kPa)	
Gr/MS [92]	Pressure	Piezoresistive	0.186 kPa^{-1}	0– < 5
PG/PET/EVA [93]	Pressure	Piezoresistive	53.99 MPa^{-1}	0.3–1000
rGrO/PET [94]	Pressure	Piezoresistive	10.39 kPa^{-1}	0–2
			0.0034 kPa^{-1}	2–200
rGrO/PDMS-	Motion/pressure	Piezoresistive	23.41 kPa^{-1} (at	0 – 30
PNWF [99]			0–3 kPa)	
			0.937 kPa^{-1} (at	
			3–30 kPa)	
rGrO/PANI WS [100]	Motion/pressure	Piezoelectric	0.042 to 0.152	0–27
			kPa^{-1}	

(continued)

Table 4.4 (continued)

Sensing material	Type	Sensing mechanism	Sensitivity	Detection range (kPa)
LAG/PDMS	Motion/pressure	Triboelectric	7.697 kPa^{-1} (at	< 1 – > 20
/PET/ITO [101]			<1 kPa)	
			0.938 kPa^{-1} (at	
			1 to 20 kPa)	
			0.136 kPa^{-1} (at	
			>20)	

Wang et al. introduced an approach for developing graphene-fabric resistive temperature sensors [112]. The process involved dip-coating a non-woven fabric (NWF) into a solution of graphene nanoplatelets (GNPs) combined with sodium alginate at an optimal concentration. The base non-woven fabric was created using a wet-spinning technique with calcium alginate fibres. The study investigated the mechanical properties of the fabric both before and after the application of the GNPs/ sodium alginate mixture. Numerical results indicated that the thermal decomposition threshold and tensile stress of the mixture increased from 230 °C to 240 °C and 99 MPa to 135 MPa, respectively. The sensor exhibited a negative temperature coefficient of resistance (TCR), as temperature sensing experiments revealed a sensitivity of 1.34% °C 1 within the 20–45 °C temperature range. It could also detect temperature changes as small as 0.1 °C and respond within 26.3 seconds. Humidity had a negligible impact on performance, with resistance changing by only 1.36% even under extreme humidity conditions (20%—100%). This variation originated from ionic conduction and moisture-induced swelling of the polymer. At higher humidity levels, the production of hydronium ions (H_3O^+) from the ionisation of water molecules on the sensor's surface decreased resistance due to ionic conductivity. However, the water caused the calcium alginate non-woven fabric to swell, increasing the separation between graphene layers and thus reducing the sensor's conductivity due to poorer graphene connectivity. These findings demonstrate that graphene has a positive impact on overall quality.

In a related study by Kun et al. [113], a LIG-based sensor was produced through direct CO_2 laser treatment of a PI film. Hydrosol and double-sided tapes were attached to an acrylic board to increase precision. Later, the hydrosol tape was dissolved in water, enabling the safe removal of the sensor from the acrylic board. The temperature responsiveness of this sensor was examined within a range of 30 °C to 40 °C, which corresponds to the threshold of human body temperature. The sensor demonstrated – 0.04145% $°C^{-1}$ sensitivity. However, it is worth noting that the non-identical thermal coefficients of LIG and PI may impact the sensor's performance.

Gong et al. developed disposable temperature sensors using graphene paper [114]. The team proposed a cost-effective, accessible, and simplified fabrication process for graphene paper sensors. By writing or mask-spraying, the team applied GNRs ink directly onto a standard paper substrate. GNRs are quasi-one-dimensional derivatives of graphene and exhibit solution processability and a low defect density that are beneficial for disposable sensors [114]. The synthesis of GNRs ink involves suspending MWCNTS in a mixture containing an oxidiser and catalytic agent, specifically H_2SO_4 and $KMnO_4$. This mixture facilitated the unzipping of MWCNTS, resulting in the formation of nanoribbons. Subsequently, an unzipping procedure introduces oxygen-containing functional groups, such as hydroxyl and carboxyl, into the nanoribbons. This modification improved the stability and dispersity of the ink in polar solvents, including water. The ink application utilised a PET-based plastic mask in conjunction with metal masks, followed by ambient drying of the prototypes and encapsulation using Scotch tape. The performance evaluation of the sensor revealed a negative average TCR of $-1.27\%\ ^\circ C^{-1}$ within the temperature range of 30 °C to 80 °C.

The nanoscale lateral dimensions of the GNRs endow them with a specific bandgap and local traps within the energy band, thereby enhancing thermally activated carrier transport. This process was reported to facilitate the release of carriers localised in the energy band traps into the extended band and intrinsic excitation of free carriers, as well as increase the sensor's temperature sensitivity. The developed sensor was reported to have several advantages, including a response and recovery time of 0.5 s, a resolution of up to 0.2 °C, and bendable properties. Besides temperature measurement, the sensor was reported to monitor physiological parameters, such as respiration rate and human touch, highlighting its potential versatility in various applications [114].

In 2018, Liu et al. investigated the use of rGrO to develop temperature sensors with potential applications in robotic skin and the Internet of Things (IoT) [15]. This study involved the integration of rGrO as a sensing layer sandwiched between a layer of high-temperature transparent tape, an insulator, and a PET substrate. The researchers fabricated the sensor and conducted comparative analyses with counterparts using SWCNTs and MWCNTs for the sensing layer. The fabrication methodology involved plasma etching to create irregular microstructures on the PET surface, thereby enhancing the adhesion of the carbon-based sensing layers to the substrate. A screen-printing technique was also employed to establish two conductive thin wires. The experimental results demonstrated a linear correlation between temperature variation and the resistance response of the rGrO and MWCNT-based sensors, in contrast to the nonlinear response observed in the SWCNT-based sensors. The rGrO-based sensor exhibited better sensitivity, quantified as $0.6345\%\ ^\circ C^{-1}$. The rGrO sensor was also reported to exhibit stable resistance under various pressures and deformations. This durability was achieved by reducing the interlayer spacing of graphene, which was accomplished by applying a considerable force during the bonding process to the insulating layer, as described in the literature. Moreover, the insulating layer propagated resistance to the sensor against fluctuations in surrounding humidity and other gaseous environments, underscoring its suitability for diverse applications [15].

In the study documented in reference [115], an investigation was presented on the direct 3D printing of GNPs/PDMS to fabricate sensors that exhibit both stretchability and strain insensitivity. The production of the material involved the fabrication of graphene nanoplates (GNPs) from natural graphite by processing with H_2SO_4 and fuming nitric acid. Subsequent annealing at 1050 °C for 25 seconds facilitated the production of GNPs. These GNPs were mixed with ethyl acetate and PDMS to synthesise the nanocomposite inks. Notably, these inks retained their applicability for up to 60 days when stored at a temperature of 4 °C. During the evaluative phase, encompassing temperature ranges from 25 °C to 75 °C, the sensors demonstrated a positive TCR with a sensitivity of 0.8% C^{-1}. The sensors primarily consisted of a grid-like porous architecture, which exhibited response and recovery times ranging from 1.31 to 3.91 s and from 1.01 to 4.58 s, respectively.

Zhang et al. (2020) developed a multi-modal sensor that can detect strain and temperature. This sensor was developed by integrating PEDOT/PSS with rGrO into an aerogel structure, which was then embedded within PDMS [116]. The innovative PEDOT/PSS/rGrO aerogel was created by freeze-drying a mixture of PEDOT/PSS and GrO nanosheets assembled through chemical reduction-induced self-assembly. The aerogel was mixed with a PDMS precursor, enabling the polymer to permeate its porous structure and form a flexible strain sensor. This sensor detects both strain and temperature variations through an impedance-based approach. The sensor enables the simultaneous measurement of strain and temperature with a single device by evaluating the electrical impedance at two separate frequencies. The gauge factor for the sensor, which utilises a nanocomposite of PEDOT/PSS/rGrO/aerogel at a weight ratio of 1:30 and a total filler concentration of 2.5 mg mL^{-1}, was determined to be approximately 57.6. Furthermore, the sensor demonstrated a decrease in resistance with an increase in temperature, indicating a negative TCR with a sensitivity of –1.69% $°C^{-1}$ within a temperature range of 30–50 °C. Figs. 4.10, 4.11, 4.12 and Table 4.5 present different types of graphene-based temperature sensors and their features.

4.5 Summary

This chapter presented the potential application of graphene-based wearable sensors for detecting biophysical signals. The graphene-based biophysical sensor can be categorised into three groups: electrophysiological (e.g., ECG, EMG, EEG, and EOG), kinematic (including acoustic, pressure, and motion), and body temperature, based on the type of signal. Summary tables were included to outline the distinct characteristics of biophysical sensors.

The next chapter will discuss the associated challenges and propose a future direction for graphene-based non-invasive biomedical sensors.

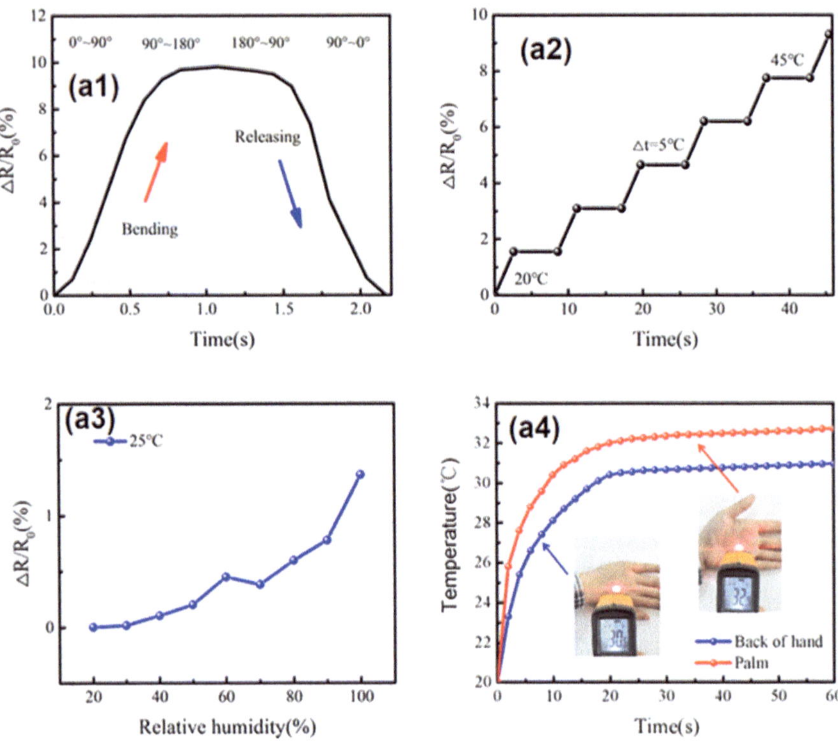

Fig. 4.10 **a1** Rnormalisedchange rate curve of FTS with bending strain (25 °C). **a2** Modification curve of FTS for bending strain-temperature. **a3** Resistance change rate curves of FTS with different humidity (25 °C). **a4** Accurately monitor the temperature difference between the palm and the back of the hand by FTS. (Reproduced from [112] with permission from Springer Nature)

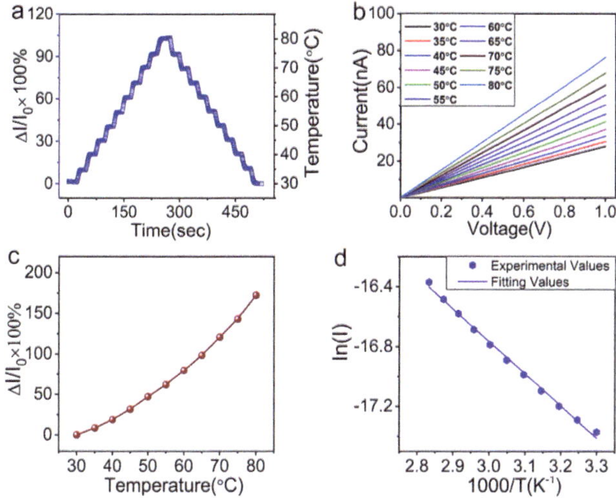

Fig. 4.11 **a** Step-like curve with the variation of temperature of the GNRs sensor. The right y-coordinate corresponds to the temperature of the normalised current change. **b** I–V curves of the sensor from 30 to 80 °C. **c** Plot of normalised current change versus temperature. **d** showing linear relationship between ln(I) and 1000/T. (Reproduced from [114] with permission from RSC)

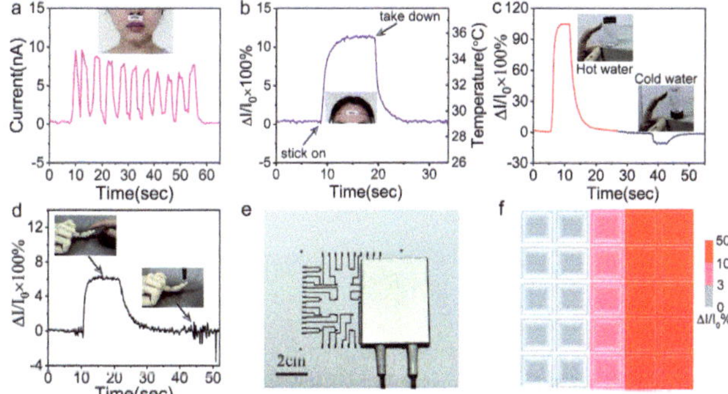

Fig. 4.12 **a** A GNR sensor was attached below the nose to monitor human respiratory rate in a calm state through sensing the temperature difference of breathing gas. **b** Measurement of human body temperature by the GNRs sensor adhered to the forehead. The GNR sensor is directly attached to the artificial hand for **c** sensing water with different temperatures and **d** identifying human touch. **e** Photo of the GNRs sensor array covered by a hot plate. **f** Temperature distributions measured by the GNRs sensor array. (Reproduced from [114] with permission from RSC)

Table 4.5 Graphene-based temperature sensing materials and their properties

Sensing material	Operating temperature range (°C)	Sensitivity	Response time (sec)
SA nanofibril /GrO/PAM [111]	25–65	$2.0 - 2.7\%$ °C^{-1} with strain from 0–200%	~2.5
GNPs/NWF [112]	20–45	1.34% °C^{-1}	~26.3
LIG/CO_2-polyimide [113]	30–40	−0.04145% °C^{-1}	30
GNRs/paper/PET [114]	30–80	Relative current change rate 172%	~0.5
rGrO/SWNTs /MWCNTs/PET [15]	30–100	0.6345% °C^{-1}	~1.2
GNPs/PDMS [115]	25–75	0.8% °C^{-1}	~1.32–3.91
PEDOT/PSS/rGrO/ aerogel [116]	30–50	−1.69% °C^{-1}	U[4.1]

[1]U = Unknown

References

1. Z. Chu, W. Zhang, Q. You, X. Yao, T. Liu, G. Liu, G. Zhang, X. Gu, Z. Ma, W. Jin, A separation-sensing membrane performing precise real-time serum analysis during blood drawing. Angew. Chem. **132**(42), 18860–18867 (2020)
2. Y. Wang, H. Haick, S. Guo, C. Wang, S. Lee, T. Yokota, T. Someya, Skin bioelectronics towards long-term, continuous health monitoring. Chem. Soc. Rev. **51**(9), 3759–3793 (2022)
3. Ahmad Tarar, A., Mohammad, U., K. Srivastava, S.: Wearable skin sensors and their challenges: A review of transdermal, optical, and mechanical sensors. Biosensors **10**(6), 56 (2020)
4. Li, Z., Zhang, S., Chen, Y., Ling, H., Zhao, L., Luo, G., Wang, X., Hartel, M., Liu, H., Xue, Y., Haghniaz, R., Lee, K., Sun, W., Kim, H., Lee, J., Zhao, Y., Zhao, Y., Emaminejad, S., Aha- dian, S.: Gelatin methacryloyl-based tactile sensors for medical wearables. Advanced Functional Materials **30** (2020) https://doi.org/10.1002/adfm.202003601
5. S. Li, Z. Ma, Z. Cao, L. Pan, Y. Shi, Advanced wearable microfluidic sensors for healthcare monitoring. Small **16**(9), 1903822 (2020)
6. Y. Lee, J. Kim, B. Jang, S. Kim, B.K. Sharma, J.-H. Kim, J.-H. Ahn, Graphene-based stretchable/wearable self-powered touch sensor. Nano Energy **62**, 259–267 (2019)
7. R. Ghaffari, J.A. Rogers, T.R. Ray, Recent progress, challenges, and opportunities for wearable biochemical sensors for sweat analysis. Sens. Actuators, B Chem. **332**, 129447 (2021)
8. Jayathilaka, W.A.D.M., Qi, K., Qin, Y., Chinnappan, A., Serrano-Garc´ıa, W., Baskar, C., Wang, H., He, J., Cui, S., Thomas, S.W., *et al.*: Significance of nanomaterials in wearables: a review on wearable actuators and sensors. Advanced Materials **31**(7), 1805921 (2019)
9. S.P. Sreenilayam, I.U. Ahad, V. Nicolosi, V.A. Garzon, D. Brabazon, Advanced materials of printed wearables for physiological parameter monitoring. Mater. Today **32**, 147–177 (2020)
10. D.H. Ho, P. Hong, J.T. Han, S.-Y. Kim, S.J. Kwon, J.H. Cho, 3d-printed sugar scaffold for high-precision and highly sensitive active and passive wearable sensors. Advanced Science **7**(1), 1902521 (2020)
11. S. Wang, Y. Jiang, H. Tai, B. Liu, Z. Duan, Z. Yuan, H. Pan, G. Xie, X. Du, Y. Su, An integrated flexible self-powered wearable respiration sensor. Nano Energy **63**, 103829 (2019)

12. X. Mo, H. Zhou, W. Li, Z. Xu, J. Duan, L. Huang, B. Hu, J. Zhou, Piezoelectrets for wearable energy harvesters and sensors. Nano Energy **65**, 104033 (2019)
13. G. Chen, H. Zhang, H. Wang, F. Wang, Immune tolerance induced by immune-homeostatic particles. Engineered Regeneration **2**, 133–136 (2021)
14. H. Zhang, R. He, Y. Niu, F. Han, J. Li, X. Zhang, F. Xu, Graphene-enabled wearable sensors for healthcare monitoring. Biosens. Bioelectron. **197**, 113777 (2022)
15. Y. Liu, H. Wang, W. Zhao, M. Zhang, H. Qin, Y. Xie, Flexible, stretchable sensors for wearable health monitoring: sensing mechanisms, materials, fabrication strategies and features. Sensors **18**(2), 645 (2018)
16. C. Wang, K. Xia, H. Wang, X. Liang, Z. Yin, Y. Zhang, Advanced carbon for flexible and wearable electronics. Adv. Mater. **31**(9), 1801072 (2019)
17. H. Huang, S. Su, N. Wu, H. Wan, S. Wan, H. Bi, L. Sun, Graphene-based sensors for human health monitoring. Front. Chem. **7**, 399 (2019)
18. M. Wang, Y. Yang, W. Gao, Laser-engraved graphene for flexible and wearable electronics. Trends in Chemistry **3**(11), 969–981 (2021)
19. N. Baig, I. Kammakakam, W. Falath, Nanomaterials: A review of synthesis methods, properties, recent progress, and challenges. Materials advances **2**(6), 1821–1871 (2021)
20. P. Suvarnaphaet, S. Pechprasarn, Graphene-based materials for biosensors: a review. Sensors **17**(10), 2161 (2017)
21. C.I. Justino, A.R. Gomes, A.C. Freitas, A.C. Duarte, T.A. Rocha-Santos, Graphene based sensors and biosensors. TrAC, Trends Anal. Chem. **91**, 53–66 (2017)
22. S. Szunerits, R. Boukherroub, Graphene-based biosensors. Interface focus **8**(3), 20160132 (2018)
23. S. Wan, Z. Zhu, K. Yin, S. Su, H. Bi, T. Xu, H. Zhang, Z. Shi, L. He, L. Sun, A highly skin-conformal and biodegradable graphene-based strain sensor. Small Methods **2**, 1700374 (2018). https://doi.org/10.1002/smtd.201700374
24. Miao, P., Wang, J., Zhang, C., Sun, M., Cheng, S., Liu, H.: Graphene nanostructure-based tactile sensors for electronic skin applications. Nano-Micro Letters **11**, 71 (2019) https://doi.org/10.1007/s40820-019-0302-0
25. J. Kim, M. Kim, M.-S. Lee, K. Kim, S. Ji, Y.-T. Kim, J. Park, K. Na, K.-H. Bae, H. Kyun Kim et al., Wearable smart sensor systems integrated on soft contact lenses for wireless ocular diagnostics. Nat. Commun. **8**(1), 14997 (2017)
26. Q. Zhang, L. Tan, Y. Chen, T. Zhang, W. Wang, Z. Liu, L. Fu, Human-like sensing and reflexes of graphene-based films. Advanced Science **3**(12), 1600130 (2016)
27. Choi, C., Lee, Y., Cho, K.W., Koo, J., Kim, D.-H.: Wearable and implantable soft bioelectronics using two-dimensional materials. Accounts of Chemical Research **52** (2018) https://doi.org/ https://doi.org/10.1021/acs.accounts.8b00491
28. A. Singh, A. Ahmed, A. Sharma, S. Arya, Graphene and its derivatives: synthesis and application in the electrochemical detection of analytes in sweat. Biosensors **12**(10), 910 (2022)
29. M. Zhu, H. Wang, S. Li, X. Liang, M. Zhang, X. Dai, Y. Zhang, Flexible electrodes for in vivo and in vitro electrophysiological signal recording. Adv. Healthcare Mater. **10**(17), 2100646 (2021)
30. Y. Li, Y. Li, S. Wu, X. Wu, J. Shu, Laser-scribed graphene for human health monitoring: From biophysical sensing to biochemical sensing. Nanomaterials **14**(11), 942 (2024)
31. Y. Qiao, X. Li, T. Hirtz, G. Deng, Y. Wei, M. Li, S. Ji, Q. Wu, J. Jian, F. Wu et al., Graphene-based wearable sensors. Nanoscale **11**(41), 18923–18945 (2019)
32. S. Kabiri Ameri, R. Ho, H. Jang, L. Tao, Y. Wang, L. Wang, D.M. Schnyer, D. Akinwande, N. Lu, Graphene electronic tattoo sensors. ACS nano **11**(8), 7634–7641 (2017)
33. S. Ramasamy, A. Balan, Wearable sensors for ecg measurement: a review. Sens. Rev. **38**(4), 412–419 (2018)
34. Golparvar, A.J., Yapici, M.K.: Wearable graphene textile-enabled eog sensing. In: 2017 IEEE SENSORS, pp. 1–3 (2017), IEEE

35. C.-Y. Huang, C.-W. Chiu, Facile fabrication of a stretchable and flexible nanofiber carbon film-sensing electrode by electrospinning and its application in smart clothing for ecg and emg monitoring. ACS Appl. Electron. Mater. **3**(2), 676–686 (2021)
36. G. Murastov, E. Bogatova, K. Brazovskiy, I. Amin, A. Lipovka, E. Dogadina, A. Cherepnyov, A. Ananyeva, E. Plotnikov, V. Ryabov et al., Flexible and water-stable graphene-based electrodes for long-term use in bioelectronics. Biosens. Bioelectron. **166**, 112426 (2020)
37. M.K. Yapici, T.E. Alkhidir, Intelligent medical garments with graphene-functionalized smart-cloth ecg sensors. Sensors **17**(4), 875 (2017)
38. T. Kim, J. Park, J. Sohn, D. Cho, S. Jeon, Bioinspired, highly stretchable, and conductive dry adhesives based on 1d–2d hybrid carbon nanocomposites for all-in-one ecg electrodes. ACS Nano **10**(4), 4770–4778 (2016)
39. D.W. Kim, S. Baik, H. Min, S. Chun, H.J. Lee, K.H. Kim, J.Y. Lee, C. Pang, Highly permeable skin patch with conductive hierarchical architectures inspired by amphibians and octopi for omnidirectionally enhanced wet adhesion. Adv. Func. Mater. **29**(13), 1807614 (2019)
40. J. Qiu, T. Yu, W. Zhang, Z. Zhao, Y. Zhang, G. Ye, Y. Zhao, X. Du, X. Liu, L. Yang et al., A bioinspired, durable, and nondisposable transparent graphene skin electrode for electrophysiological signal detection. ACS Mater. Lett. **2**(8), 999–1007 (2020)
41. X. Xu, M. Luo, P. He, J. Yang, Washable and flexible screen printed graphene electrode on textiles for wearable healthcare monitoring. J. Phys. D Appl. Phys. **53**(12), 125402 (2020)
42. Q. Wang, S. Ling, X. Liang, H. Wang, H. Lu, Y. Zhang, Self-healable multifunctional electronic tattoos based on silk and graphene. Adv. Func. Mater. **29**(16), 1808695 (2019)
43. X. Pan, Q. Wang, P. He, K. Liu, Y. Ni, L. Chen, X. Ouyang, L. Huang, H. Wang, S. Xu, A bionic tactile plastic hydrogel-based electronic skin constructed by a nerve-like nanonetwork combining stretchable, compliant, and self-healing properties. Chem. Eng. J. **379**, 122271 (2020)
44. A.S. Prasad, M. Jayaram et al., Fabrication of gnr electrode for ecg signal acquisition. IEEE Sens. Lett. **5**(9), 1–4 (2021)
45. P.-Y. Huang, C.-Y. Huang, J.-W. Li, S.-Y. Shen, C.-C. Cheng, C.-W. Chiu, R.-J. Jeng, J.-J. Lin, Immobilization of air-stable copper nanoparticles on graphene oxide flexible hybrid films for smart clothes. Polymers **14**(2), 237 (2022)
46. X. Du, W. Jiang, Y. Zhang, J. Qiu, Y. Zhao, Q. Tan, S. Qi, G. Ye, W. Zhang, N. Liu, Transparent and stretchable graphene electrode by intercalation doping for epidermal electrophysiology. ACS Appl. Mater. Interfaces. **12**(50), 56361–56371 (2020)
47. Suvarnaphaet, P., Sasivimolkul, S., Sukkasem, C., Pukesamsombut, D., Tanadchangsaeng, N., Boonyagul, S., Pechprasarn, S.: Biodegradable electrode patch made of graphene/pha for ecg detecting applications. In: 2019 12th Biomedical Engineering International Conference (BMEiCON), pp. 1–5 (2019). https://doi.org/10.1109/BMEiCON47515.2019.8990243 . IEEE
48. W.S. Hummers Jr., R.E. Offeman, Preparation of graphitic oxide. J. Am. Chem. Soc. **80**(6), 1339–1339 (1958)
49. T.-R. Cui, D. Li, X.-R. Huang, A.-Z. Yan, Y. Dong, J.-D. Xu, Y.-Z. Guo, Y. Wang, Z.-K. Chen, W.-C. Shao et al., Graphene-based flexible electrode for electrocardiogram signal monitoring. Appl. Sci. **12**(9), 4526 (2022)
50. M.A. Zahed, S.C. Barman, M. Sharifuzzaman, S. Zhang, H. Yoon, C. Park, S.H. Yoon, J.Y. Park, Polyaziridine-encapsulated phosphorene-incorporated flexible 3d porous graphene for multimodal sensing and energy storage applications. Adv. Func. Mater. **31**(25), 2009018 (2021)
51. Zhang, S., Chhetry, A., Zahed, M.A., Sharma, S., Park, C., Yoon, S., Park, J.Y.: On-skin ultrathin and stretchable multifunctional sensor for smart healthcare wearables. npj Flexible Electronics **6**(1), 11 (2022)
52. Maher, A., Mahmoud, M., Said, A.: Synthesis of transparent bio-electrodes for biophysiological measurements based on modified graphene oxide. Nanotechnology **33** (2021) https://doi.org/ https://doi.org/10.1088/1361-6528/ac2e23

53. J. Song, Y. Kim, K. Kang, S. Lee, M. Shin, D. Son, Stretchable and self-healable graphene–polymer conductive composite for wearable EMG sensor. Polymers **14**(18), 3766 (2022)
54. J.-W. Li, J.C.-M. Lee, K.-C. Chuang, C.-W. Chiu, Photocured, highly flexible, and stretchable 3d-printed graphene/polymer nanocomposites for electrocardiography and elec- tromyography smart clothing. Prog. Org. Coat. **176**, 107378 (2023)
55. B. Stephens-Fripp, V. Sencadas, R. Mutlu, G. Alici, Reusable flexible concentric electrodes coated with a conductive graphene ink for electrotactile stimulation. Front. Bioeng. Biotechnol. **6**, 179 (2018)
56. A. Del Vecchio, A. Holobar, D. Falla, F. Felici, R. Enoka, D. Farina, Tutorial: Analy- sis of motor unit discharge characteristics from high-density surface emg signals. J. Electromyogr. Kinesiol. **53**, 102426 (2020)
57. I. Yun, J. Jeung, H. Lim, J. Kang, S. Lee, S. Park, S. Seong, S. Park, K. Cho, Y. Chung, Stable bioelectric signal acquisition using an enlarged surface-area flexible skin electrode. ACS Appl. Electron. Mater. **3**(4), 1842–1851 (2021)
58. Ozturk, O., Yapici, M.K.: Muscular activity monitoring and surface electromyography (semg) with graphene textiles. In: 2019 IEEE SENSORS, pp. 1–4 (2019). IEEE
59. Ozturk, O., Golparvar, A., Yapici, M.K.: Smart armband with graphene textile electrodes for emg-based muscle fatigue monitoring. In: 2021 Ieee Sensors, pp. 1–4 (2021). IEEE
60. P.S. Das, S.H. Park, K.Y. Baik, J.W. Lee, J.Y. Park, Thermally reduced graphene oxide-nylon membrane based epidermal sensor using vacuum filtration for wearable electro-physiological signals and human motion monitoring. Carbon **158**, 386–393 (2020)
61. J.-W. Li, H.-F. Chen, Y.-Z. Liu, J.-H. Wang, M.-C. Lu, C.-W. Chiu, Photocurable 3d-printed agnps/graphene/polymer nanocomposites with high flexibility and stretchability for ecg and emg smart clothing. Chem. Eng. J. **484**, 149452 (2024)
62. Y. Pan, X. Cui, D. Song, W. Hu, X. Lin, N. Liu, A stretchable and sweat-adhesive 3d graphene eutectogel electrode for emg monitoring. ACS Appl. Nano Mater. **7**(10), 12064–12071 (2024)
63. Islam, M.R., Afroj, S., Beach, C., Islam, M.H., Parraman, C., Abdelkader, A., Casson, A.J., Novoselov, K.S., Karim, N.: Fully printed and multifunctional graphene-based wearable e-textiles for personalized healthcare applications. IScience **25**(3) (2022)
64. L. Shao, Y. Guo, W. Liu, T. Sun, D. Wei, A flexible dry electroencephalogram electrode based on graphene materials. Mater. Res. Exp. **6**(8), 085619 (2019)
65. Masvidal-Codina, E., Illa, X., Dasilva, M., Calia, A.B., Dragojevi´c, T., Vidal-Rosas, E.E., Prats- Alfonso, E., Mart´ınez-Aguilar, J., Cruz, J.M., Garcia-Cortadella, R., *et al.*: High-resolution mapping of infraslow cortical brain activity enabled by graphene microtransistors. Nature materials **18**(3), 280–288 (2019)
66. A. Golparvar, O. Ozturk, M.K. Yapici, Gel-free wearable electroencephalography (EEG) with soft graphene textiles, in *2021 IEEE Sensors*(IEEE, 2021), pp. 1–4
67. Z. Li, W. Guo, Y. Huang, K. Zhu, H. Yi, H. Wu, On skin graphene electrodes for large area electrophysiological monitoring and human-machine interfaces. Carbon **164**, 164–170 (2020)
68. A.J. Golparvar, M.K. Yapici, Electrooculography by wearable graphene textiles. IEEE Sens. J. **18**(21), 8971–8978 (2018)
69. A.J. Golparvar, M.K. Yapici, Toward graphene textiles in wearable eye tracking systems for human–machine interaction. Beilstein J. Nanotechnol. **12**(1), 180–189 (2021)
70. C. Beach, N. Karim, A.J. Casson, A graphene-based sleep mask for comfortable wearable eye tracking, in *2019 41st Annual International Conference of the IEEE Engineering in Medicine and Biology Society (EMBC)* (IEEE, 2019), pp. 6693–6696
71. L.-Q. Tao, D.-Y. Wang, H. Tian, Z.-Y. Ju, Y. Liu, Y. Pang, Y.-Q. Chen, Y. Yang, T.-L. Ren, Self-adapted and tunable graphene strain sensors for detecting both subtle and large human motions. Nanoscale **9**(24), 8266–8273 (2017)
72. Y. Wei, Y. Qiao, G. Jiang, Y. Wang, F. Wang, M. Li, Y. Zhao, Y. Tian, G. Gou, S. Tan et al., A wearable skinlike ultra-sensitive artificial graphene throat. ACS Nano **13**(8), 8639–8647 (2019)
73. Q. Xing, S. Li, X. Fan, A. Bian, A., S.-J. Cao, C. Li, Influential factors on thermoacoustic efficiency of multilayered graphene film loudspeakers for optimal design. J. Appl. Phys. **122**(12) (2017)

74. W.H. Li, N.T.C. Tan, Z.L. Ngoh, J.J. Yu, Y. Tan, E.H.T. Teo, Experimental and simulation analysis on impact of three-dimensional graphene material properties on thermo-acoustic sound generation, in *OCEANS 2024-Singapore* (IEEE, 2024), pp. 1–7
75. W. Hou, Y. Wei, Y. Wang, S. Duan, Z. Guo, H. Tian, Y. Yang, T.-L. Ren, A large-scale and low-cost thermoacoustic loudspeaker based on three-dimensional graphene foam. ACS Appl. Mater. Interfaces. **16**(18), 23544–23552 (2024)
76. L.-Q. Tao, H. Tian, Y. Liu, Z.-Y. Ju, Y. Pang, Y.-Q. Chen, D.-Y. Wang, X.-G. Tian, J.-C. Yan, N.-Q. Deng, Y. Yang, T. Ren, An intelligent artificial throat with sound-sensing ability based on laser induced graphene. Nat. Commun. **8**, 14579 (2017). https://doi.org/10.1038/ncomms14579
77. H. Yang, T. Xue, F. Li, W. Liu, Y. Song, Graphene: diversified flexible 2d material for wearable vital signs monitoring. Adv. Mater. Technol. **4**(2), 1800574 (2019)
78. L. Lv, P. Zhang, T. Xu, L. Qu, Ultrasensitive pressure sensor based on an ultralight sparkling graphene block. ACS Appl. Mater. Interfaces. **9**(27), 22885–22892 (2017)
79. X. Liu, D. Liu, J.-H. Lee, Q. Zheng, X. Du, X. Zhang, H. Xu, Z. Wang, Y. Wu, X. Shen et al., Spider-web-inspired stretchable graphene woven fabric for highly sensitive, transparent, wearable strain sensors. ACS Appl. Mater. Interfaces. **11**(2), 2282–2294 (2018)
80. Y. Pang, K. Zhang, Z. Yang, S. Jiang, Z. Ju, Y. Li, X. Wang, D. Wang, M. Jian, Y. Zhang et al., Epidermis microstructure inspired graphene pressure sensor with random distributed spinosum for high sensitivity and large linearity. ACS Nano **12**(3), 2346–2354 (2018)
81. Y. Pang, H. Tian, L. Tao, Y. Li, X. Wang, N. Deng, Y. Yang, T.-L. Ren, Flexible, highly sensitive, and wearable pressure and strain sensors with graphene porous network structure. ACS Appl. Mater. Interfaces. **8**(40), 26458–26462 (2016)
82. Z. Chen, Z. Wang, X. Li, Y. Lin, N. Luo, M. Long, N. Zhao, J.-B. Xu, Flexible piezoelectric-induced pressure sensors for static measurements based on nanowires/graphene heterostructures. ACS Nano **11**(5), 4507–4513 (2017)
83. M.B. Coskun, L. Qiu, M.S. Arefin, A. Neild, M. Yuce, D. Li, T. Alan, Detecting subtle vibrations using graphene-based cellular elastomers. ACS Appl. Mater. Interfaces. **9**(13), 11345–11349 (2017)
84. C.-B. Huang, S. Witomska, A. Aliprandi, M.-A. Stoeckel, M. Bonini, A. Ciesielski, P. Samor‘ı, Molecule–graphene hybrid materials with tunable mechanoresponse: highly sensitive pressure sensors for health monitoring. Adv. Mater. **31**(1), 1804600 (2019)
85. H. Kou, L. Zhang, Q. Tan, G. Liu, W. Lv, F. Lu, H. Dong, J. Xiong, Wireless flexible pressure sensor based on micro-patterned graphene/pdms composite. Sens. Actuators A **277**, 150–156 (2018)
86. J. Shi, L. Wang, Z. Dai, L. Zhao, M. Du, H. Li, Y. Fang, Multiscale hierarchical design of a flexible piezoresistive pressure sensor with high sensitivity and wide linearity range. Small **14**(27), 1800819 (2018)
87. W. Chen, X. Gui, B. Liang, R. Yang, Y. Zheng, C. Zhao, X. Li, H. Zhu, Z. Tang, Structural engineering for high sensitivity, ultrathin pressure sensors based on wrinkled graphene and anodic aluminum oxide membrane. ACS Appl. Mater. Interfaces. **9**(28), 24111–24117 (2017)
88. Z. Lou, S. Chen, L. Wang, K. Jiang, G. Shen, An ultra-sensitive and rapid response speed graphene pressure sensors for electronic skin and health monitoring. Nano Energy **23**, 7–14 (2016)
89. N. Luo, Y. Huang, J. Liu, S. Chen, C. Wong, Hollow-structured graphene–silicone- composite-based piezoresistive sensors: decoupled property tuning and bending reliability. Adv. Mater. **29** (2017). https://doi.org/10.1002/adma.201702675
90. L. Zhang, H. Li, X. Lai, T. Gao, J. Yang, X. Zeng, Thiolated graphene@ polyester fabric-based multilayer piezoresistive pressure sensors for detecting human motion. ACS Appl. Mater. Interfaces. **10**(48), 41784–41792 (2018)
91. A. Tewari, S. Gandla, S. Bohm, C.R. McNeill, D. Gupta, Highly exfoliated mwnt–rgo ink-wrapped polyurethane foam for piezoresistive pressure sensor applications. ACS Appl. Mater. Interfaces. **10**(6), 5185–5195 (2018)

92. F. Zhang, K. Yang, Z. Pei, Y. Wu, S. Sang, Q. Zhang, H. Jiao, A highly accurate flexible sensor system for human blood pressure and heart rate monitoring based on graphene/sponge. RSC Adv. **12**(4), 2391–2398 (2022)
93. Y. Peng, J. Zhou, X. Song, K. Pang, A. Samy, Z. Hao, J. Wang, A flexible pressure sensor with ink printed porous graphene for continuous cardiovascular status monitoring. Sensors **21**(2), 485 (2021)
94. Z. Yue, X. Ye, S. Liu, Y. Zhu, H. Jiang, Z. Wan, Y. Lin, C. Jia, Towards ultra-wide operation range and high sensitivity: graphene film based pressure sensors for fingertips. Biosens. Bioelectron. **139**, 111296 (2019)
95. H. Wang, X. Xu, J. Li, L. Lin, L. Sun, X. Sun, S. Zhao, C. Tan, C. Chen, W. Dang et al., Surface monocrystallization of copper foil for fast growth of large single-crystal graphene under free molecular flow. Adv. Mater. **28**(40), 8968–8974 (2016)
96. J. Shi, X. Li, H. Cheng, Z. Liu, L. Zhao, T. Yang, Z. Dai, Z. Cheng, E. Shi, L. Yang et al., Graphene reinforced carbon nanotube networks for wearable strain sensors. Adv. Func. Mater. **26**(13), 2078–2084 (2016)
97. Y. Qiao, Y. Wang, H. Tian, M. Li, J. Jian, Y. Wei, Y. Tian, D.-Y. Wang, Y. Pang, X. Geng et al., Multilayer graphene epidermal electronic skin. ACS Nano **12**(9), 8839–8846 (2018)
98. L.-Q. Tao, K.-N. Zhang, H. Tian, Y. Liu, D.-Y. Wang, Y.-Q. Chen, Y. Yang, T.- L Ren, Graphene-paper pressure sensor for detecting human motions. ACS Nano **11**(9), 8790–8795 (2017)
99. Y. Lu, M. Tian, X. Sun, N. Pan, F. Chen, S. Zhu, X. Zhang, S. Chen, Highly sensi- tive wearable 3d piezoresistive pressure sensors based on graphene coated isotropic non-woven substrate. Compos. A Appl. Sci. Manuf. **117**, 202–210 (2019)
100. G. Ge, Y. Cai, Q. Dong, Y. Zhang, J. Shao, W. Huang, X. Dong, A flexible pressure sensor based on rgo/polyaniline wrapped sponge with tunable sensitivity for human motion detection. Nanoscale **10**(21), 10033–10040 (2018)
101. P.S. Das, A. Chhetry, P. Maharjan, M.S. Rasel, J.Y. Park, A laser ablated graphene-based flexible self-powered pressure sensor for human gestures and finger pulse monitoring. Nano Res. **12**, 1789–1795 (2019)
102. D.H. Ho, Q. Sun, S.Y. Kim, J.T. Han, D.H. Kim, J.H. Cho, Stretchable and multimodal all graphene electronic skin. Adv. Mater. **28**(13), 2601–2608 (2016)
103. T.Q. Trung, S. Ramasundaram, S.W. Hong, N.-E. Lee, Flexible and transparent nanocomposite of reduced graphene oxide and p (vdf-trfe) copolymer for high thermal responsivity in a field-effect transistor. Adv. Func. Mater. **24**(22), 3438–3445 (2014)
104. T.Q. Trung, S. Ramasundaram, B.-U. Hwang, N.-E. Lee, An all-elastomeric transparent and stretchable temperature sensor for body-attachable wearable electronics. Adv. Mater. **28**(3), 502–509 (2016)
105. J. Zhang, G. Liao, S. Jin, D. Cao, Q. Wei, H. Lu, J. Yu, X. Cai, S. Tan, Y. Xiao et al., All-fiber-optic temperature sensor based on reduced graphene oxide. Laser Phys. Lett. **11**(3), 035901 (2014)
106. J. Zhu, C.M. Andres, J. Xu, A. Ramamoorthy, T. Tsotsis, N.A. Kotov, Pseudonegative thermal expansion and the state of water in graphene oxide layered assemblies. ACS Nano **6**(9), 8357–8365 (2012)
107. T.Q. Trung, H.S. Le, T.M.L. Dang, S. Ju, S.Y. Park, N.-E. Lee, Freestanding, fiber-based, wearable temperature sensor with tunable thermal index for healthcare monitoring. Adv. Healthcare Mater. **7**(12), 1800074 (2018)
108. X. Zhao, Y. Long, T. Yang, J. Li, H. Zhu, Simultaneous high sensitivity sensing of temper- ature and humidity with graphene woven fabrics. ACS Appl. Mater. Interfaces. **9**(35), 30171–30176 (2017)
109. C. Yan, J. Wang, P.S. Lee, Stretchable graphene thermistor with tunable thermal index. ACS Nano **9**(2), 2130–2137 (2015)
110. D. Kong, L.T. Le, Y. Li, J.L. Zunino, W. Lee, Temperature-dependent electrical properties of graphene inkjet-printed on flexible materials. Langmuir **28**(37), 13467–13472 (2012)

111. W. Hou, Z. Luan, D. Xie, X. Zhang, T. Yu, K. Sui, High performance dual strain- temperature sensor based on alginate nanofibril/graphene oxide/polyacrylamide nanocompos- ite hydrogel. Compos. Commun. **27**, 100837 (2021)
112. F. Wang, J. Jiang, F. Sun, L. Sun, T. Wang, Y. Liu, M. Li, Flexible wearable graphene/algi- nate composite non-woven fabric temperature sensor with high sensitivity and anti-interference. Cellulose **27**, 2369–2380 (2020)
113. H. Kun, L. Bin, M. Orban, Q. Donghai, Y. Hongbo, Accurate flexible temperature sensor based on laser-induced graphene material. Shock. Vib. **2021**(1), 9938010 (2021)
114. X. Gong, L. Zhang, Y. Huang, S. Wang, G. Pan, L. Li, Directly writing flexible temperature sensor with graphene nanoribbons for disposable healthcare devices. RSC Adv. **10**(37), 22222–22229 (2020)
115. Z. Wang, W. Gao, Q. Zhang, K. Zheng, J. Xu, W. Xu, E. Shang, J. Jiang, J. Zhang, Y. Liu, 3d-printed graphene/polydimethylsiloxane composites for stretchable and strain-insensitive temperature sensors. ACS Appl. Mater. Interfaces. **11**(1), 1344–1352 (2018)
116. F. Zhang, H. Hu, M. Islam, S. Peng, S. Wu, S. Lim, Y. Zhou, C.-H. Wang, Multi- modal strain and temperature sensor by hybridizing reduced graphene oxide and pedot: Pss. Compos. Sci. Technol. **187**, 107959 (2020)

Chapter 5
Challenges and Future Direction

Abstract This concluding chapter examines the dynamic and rapidly evolving field of graphene-based wearable sensors for health monitoring. Particular attention is given to non-invasive devices, which offer clear advantages in terms of safety and functionality. Graphene and its derivatives hold considerable promise for medical applications due to their remarkable physical and chemical properties. However, despite encouraging advances in laboratory research, a notable gap remains between experimental innovation and successful translation into clinical practice. This text critically discusses the key barriers impeding broader adoption, e.g., biocompatibility, scalability, and long-term stability issues. It also underscores the importance of developing cost-effective manufacturing techniques, ensuring consistent performance, and achieving robust mechanical integration, vital for commercial viability. Addressing these challenges will require a concerted, interdisciplinary effort. The report requests greater collaboration between materials scientists, engineers, and healthcare professionals and invites readers to reflect on the path forward in realising the full potential of these technologies to improve patient care and inform future public health strategies.

Keywords Graphene · Wearable sensors · Challenges · Sensor performance · Biomedical applications · Biocompatibility · Biological toxicity · Health monitoring · Scalability · Long-term stability · Non-invasive technology · Sensitivity · Multifunctionality

5.1 Introduction

Due to advances in technology and research, there has been significant interest in developing sensors and devices for health monitoring. Among invasive and non-invasive wearable sensors, non-invasive sensors are increasingly favoured in healthcare settings due to their minimal invasiveness and reduced risk while maintaining functionality and performance [1]. Due to the unique properties of graphene and its

S. Debnath et al., *Graphene in Wearable Sensors for Health Monitoring*, SpringerBriefs in Applied Sciences and Technology,
https://doi.org/10.1007/978-981-96-8850-0_5

potential applications, as well as those of its derivatives, the exploration and development of graphene-based sensors have been a focal point of extensive research. These sensors have been employed to measure vital signs and biomarkers in the human body. These developments thus enable transformative applications in clinical medicine, rehabilitative therapies, and augmented human–computer interaction systems [2].

Despite significant advancements in the development of graphene-based wearable sensors for health monitoring, a considerable gap exists between laboratory research and industrial applications. While graphene's exceptional physicochemical properties—such as high electrical conductivity and mechanical flexibility—have been extensively characterised, challenges including biocompatibility, scalability, and long-term stability remain unresolved. The successful integration of these technologies into practical applications for personal health management and disease detection is hindered by challenges such as cost-effective mass production, reliable sensing performance, seamless device integration, and mechanical durability. Overcoming these obstacles is crucial to transforming academic advancements into commercially viable solutions that benefit public health.

This chapter presents challenges and future research directions in graphene-based wearable biomedical sensors.

5.2 Challenges

5.2.1 Biocompatibility, Biological Toxicity, and Long-Term Usability

Achieving biocompatibility, minimising biological toxicity, and ensuring long-term usability of graphene-based sensors for personal health management and disease detection pose significant challenges. With the rise of electronic skin technology, researchers have developed electronic tattoos and lightweight, flexible patches that can be attached to the skin for monitoring various health conditions. Two primary concerns in this area are breathability and biocompatibility, particularly when these wearable devices are used over extended periods [3]. Special attention must be given to the biocompatibility issue, especially for sensors used in contact lenses and mouthguards that detect tears and saliva, as these applications require close contact with the user's body. As possible solutions, the sensors could be encapsulated in an antimicrobial protective layer, or improved purification methods could be used to produce high-quality sensing materials.

The potential cytotoxicity of graphene nanomaterials raises a critical biosafety concern requiring thorough evaluation before biomedical use. Simple cytotoxicity tests are insufficient as conclusive indicators of biocompatibility for clinical applications. As demonstrated in Chap. 3, the physico-chemical properties of graphene play a pivotal role in its toxicity. Therefore, when evaluating toxicity, it is crucial to assess all relevant physicochemical factors—such as size, shape, layer thickness, lateral dimensions, atomic composition, and biological factors, including concentration, exposure duration, exposure method, and impurities [4].

The production process of graphene and its derivatives affects the biocompatibility and toxicity. High-quality CVD graphene provides valuable insights into cell interactions. Most in vitro studies on cell culture presented that nanoscale CVD graphene is biocompatible with various cell lines, enhancing fibroblast adhesion and facilitating the differentiation of human mesenchymal stem cells (hMSCs) into bone cells [5]. In contrast, producing LPE graphene sheets requires toxic solvents and surfactants for the exfoliation of graphite. These chemical substances, particularly surfactants, are challenging to remove from the final graphene sheets, which limits the applicability of LPE graphene in the biomedical field [6].

GrO produced through the modified Hummers process is widely used to investigate its interactions and effects on mammalian cells in vitro and in vivo. However, the literature presented discrepancies in findings regarding the biocompatibility and cytotoxicity of GrOs and rGrOs in different experimental conditions [5]. This inconsistency can be attributed to variations in oxidation durations, types, and concentrations of oxidants used during synthesis, which result in GrOs that differ in structure, reactivity, and levels of impurities across various studies, thereby affecting cell-GrO interactions. Furthermore, the chemical oxidisers and reducing agents can introduce metallic contaminants and organic impurities. It could also affect their interactions with cells, tissues, and organs, ultimately leading to cellular damage and apoptosis. Notably, toxic reducing agents like hydrazine can cause considerable cell destruction in human mesenchymal stem cells at relatively low concentrations of rGrO ($1.0\ \mu g\ mL^{-1}$) [7]. At the same time, residual Na_2S in rGrO can trigger severe inflammatory responses in macrophages [8]. Such adverse effects help explain the inconsistencies in the findings reported in the literature. On the contrary, GrO and rGrO synthesised through green methods show promise as nanomaterials for biomedical applications due to their reduced cytotoxicity [9].

The toxicity associated with GrOs and rGO can be reduced by incorporating them into polymers to form polymer nanocomposites. This integration confines the graphene nanomaterials within the polymer matrix, covering the sharp edges of GrOs and rGOs, thereby preventing their penetration into the cytoplasm [5]. As a result, graphene nanofillers with large surface areas provide ideal sites for cell adhesion and growth. Moreover, the oxygenated groups present in GrOs can decrease the hydrophobic nature of the polymeric matrix, thereby promoting cell attachment and spreading on the polymer surface [10, 11]. Further safety assessments and additional research are necessary to confirm the biocompatibility of these polymer nanocomposites with human tissues before they are used in clinical settings.

5.2.2 Mass Production and Cost

Mass production, the cost of graphene production, and sensor fabrication methods are key issues in replacing existing materials and commercialization [12]. From a practical standpoint, achieving scalable production of consistent and reliable graphene for graphene-based sensors at an affordable cost is a significant concern [13]. Furthermore, thoroughly exploring advanced printing and roll-to- roll processing techniques is essential for the cost-effective, large-scale production of carbon-based flexible electronics, facilitating their commercialisation and real-world applications [14]. CVD is a widely used technique for producing reproducible, dependable, and stable monolayer graphene on a large scale. In the case of CVD graphene, the size of the monolayer graphene area is relatively extensive, which complicates the achievement of uniformity in graphene synthesis [15]. Furthermore, attaching graphene to a substrate complicates maintaining durability during repeated movements or washes [16]. The complexities of the transfer process and the sensitivity of monolayer graphene indicate that synthesising graphene directly on Si or SiO_2 with high quality would considerably simplify the fabrication process [15, 17].

The synthesis of monolayer graphene typically requires temperatures exceeding 1000 °C, rendering it challenging to grow high-quality graphene directly on flexible substrates such as polydimethylsiloxane (PDMS), polyethylene terephthalate (PET), and polyimide (PI). Plasma-enhanced chemical vapour deposition (PECVD) offers a promising solution by reducing the synthesis temperature to approximately 400 °C [18]. This facilitates the production of graphene on flexible substrates, improving compatibility for wearable and flexible electronic applications. Several methods demonstrate significant potential for the reproducible and reliable production of graphene. Methods such as multilayer and porous graphene fabrication, liquid-phase exfoliation (LPE), reduction of GrO, and laser scribing enable large-scale synthesis while maintaining structural integrity and performance. These methods could meet the demands of industrial applications, such as flexible sensors and energy storage systems, by providing economically viable and consistent approaches to graphene production.

Nevertheless, the fabrication of high-resolution multilayer graphene remains a considerable challenge. The considerable roughness associated with multilayer graphene complicates the fabrication process, as thermal conduction tends to widen the actual graphene patterns beyond the intended designs from laser scribing. Furthermore, the edges of laser-scribed graphene often lack the desired regularity [15]. Several factors need to be considered during the GrO reduction process, including the concentration of the GrO solution, the thickness of the film, the type of substrate, and other relevant variables.

Ensuring uniform graphene dispersion, especially when integrated into polymers such as PDMS, TPU, and Ecoflex, presents a significant challenge. Although PDMS, TPU, and Ecoflex are frequently utilised as substrates, they may impact human comfort due to their limited air permeability and low biodegradability. Additionally, their bonding strength to the human body is inadequate, and using tape to secure sensors may compromise user comfort [16]. Researchers are exploring new microstructures and composites that offer improved properties. Further investigation is required to find biodegradable materials with better adhesion and breathability. Furthermore, producing high-quality graphene and developing sensors, particularly for those employing traditional photolithography and etching techniques, could be costly [16]. Screen printing, inkjet printing, and 3D printing show promise due to their affordability, precision, speed, and potential for roll-to-roll production [19]. Through 3D printing, integrated and miniaturised sensor systems can be integrated through a single manufacturing step. Further research is required to explore graphene-based inks or 3D-printable materials with the ideal physicochemical properties for practical fabrication and optimisation of the printing processes.

5.2.3 *Microfabrication*

Integrating graphene sensors with conventional microfabrication techniques remains a challenge. Additionally, the challenge lies in effectively integrating wearable graphene sensors with signal transmission, data processing units, and power supplies to minimise size and enhance performance [13]. Ensuring precise integration between graphene components and backend circuitry through interface optimisation is crucial for improving both signal fidelity and user comfort in wearable applications [15].

5.2.4 *Power Source*

In disease management, the development of integrated multifunctional sensors capable of facilitating feedback point-of-care therapy is highlighted as a significant innovation. The challenge of this approach requires a reliable power source with energy-utilising technologies, such as photovoltaics, thermoelectrics, and radio frequency, to support the sensors intended for long-term applications [2]. The limitations of self-powered generation and energy capacity may fall short of fulfilling data processing and transmission requirements [16]. For graphene to function effectively in wearable systems that require completely autonomous and wireless setups to monitor various health metrics, it is essential to develop effective energy storage solutions or self-powering methods. Potential mechanisms for powering such sensors include triboelectricity, piezoelectricity, enzymatic, and non-enzymatic biofuel cells [12].

However, there is a growing trend in wearable sensors towards wireless power supply to improve user convenience. Near-field communication (NFC), Bluetooth, and Wi-Fi are prominent technologies utilised in wearable device platforms because of their widespread application [20–22]. However, wireless connections, particularly NFC, often require proximity to the device [23]. Therefore, advancing wireless communication technology focusing on low energy consumption and high output efficiency is vital for the growth of wearable electronics.

5.2.5 *Sensitivity, Flexibility, Stretchability, Adhesion, Durability, and Linearity*

The challenge is also to improve the sensing performance of graphene and its derivatives, which is essential for the effectiveness of wearable healthcare monitoring systems. Besides, it is also challenging to incorporate all performance parameters, such as sensitivity, stretchability, durability, and linearity, into a single sensing material. Achieving a desirable balance between high sensitivity and extensive stretchability in strain sensors is often challenging due to inherent mechanical and electrical property limitations [2]. These limitations can restrict their effectiveness in detecting both fine and extensive human movements. Long-term durability also remains a key requirement for practical use.

Additionally, achieving a balance between high sensitivity to temperature and the necessary flexibility or stretchability presents a challenge for flexible, stretchable, and bendable temperature sensors [19]. Existing literature often indicates that highly sensitive sensors are limited to low-strain ranges. Conversely, the temperature sensitivity should remain unaffected by the strain applied to the sensors in wearable applications. Achieving a wide working range with high sensitivity and a wide linear region for high stretchability remains challenging for graphene-based strain, pressure, and temperature sensors.

Moreover, products intended for practical daily use must exhibit strong mechanical stability, particularly for graphene utilised in wearable sensors. Wearable sensors typically consist of conductive networks that create pathways for signals and flexible, stretchable elastomeric substrates that provide both flexibility and mechanical strength for these conductive networks [13]. The careful selection of materials and the design of the sensor's structure are crucial for achieving high sensitivity and stretchability. The flexibility of graphene materials often does not adequately align with the standards required for wearables [23]. Discrepancies in mechanical strength frequently occur between the sensing unit and the flexible substrate. The mechanical performance of wearable sensors is primarily determined by the flexibility and stretchability of their substrates, making the choice of substrate crucial for the overall flexibility of the sensor. An alternative approach to this challenge involves leveraging local stiffness to enhance performance. By incorporating strain shielding in the specific area of the sensing module within the soft substrate, the mechanical strength in that region can be adjusted to match that of the rigid electronic components adequately [23]. This approach provides a relatively stable environment for sensor operation, even during significant body movements. It is also important to consider the adhesion between the sensor and the skin. Adequate adhesion directly influences the breathability of skin as well as the sensor's ability to monitor health conditions effectively. Also, recognising that defective graphene can experience sudden failures due to stress corrosion cracking in different environmental settings is important [13].

Furthermore, due to the complexity and specificity of human biofluids and breath, the sensing performance of biochemical sensors may not always be ideal for actual monitoring. It is important to thoroughly understand the sensing mechanisms, particularly in the context of biochemical detection, to identify and rectify associated challenges. For instance, strong acids (e.g. HNO_3, H_2SO_4, and $HClO_4$) and potent oxidants (e.g. $KClO_3$) are commonly used in the chemical manufacturing of graphene materials [23]. If these substances are not sufficiently purified, residual acids and oxidants can damage the sensors and impair their performance. Another challenge that graphene-based sensors face is the ability to consistently and accurately detect multiple signals within compact, multifunctional wearable devices. Mechanical gaps can develop during assembly due to discrepancies between rigid components and flexible substrates. Although graphene offers the advantages of being used as a biomedical sensor, its zero-gap structure results in a comparatively low on/off ratio relative to FETs, limiting its utility in specific biomedical applications. This limitation has led to approaches such as functionalising organic molecules and exploring strain-engineered lattice distortions and spintronics [24]. Moreover, the lack of selectivity towards target analytes and the excessive sensitivity of sensors to external stimuli require surface modifications or the addition of selective layers, such as metal-organic frameworks, as suggested by Tan et al. [25].

Graphene-based sensors face several challenges, including the risk of surface contamination from substances such as hydrocarbons and the need to adapt to fluctuating humidity, temperature, and mechanical conditions, which can adversely affect sensor performance. Moreover, the detection of body fluids is hindered by factors like low biomarker concentrations, evaporation, and contamination. It is essential to find an optimal balance between these competing performance characteristics and minimise interference among various stimuli, while also addressing the effects of signal drift for accurate signal detection. Encapsulation or the use of a hydrophobic substrate can effectively prevent exposure to H_2O and O_2. This enhances the device's performance [26]. Developing encapsulation materials or cryptologic algorithms might also address these issues [27, 28]. Employing microfluidic designs for pre-concentration, separate storage, on site sampling, and highly sensitive sensing technologies could address these challenges [23].

It is challenging to evaluate the quality of these sensors objectively, as currently, there are no standardised methods for evaluating the performance of graphene-based sensors. Establishing standards for excellent performance while continuously aiming for improvements is important. In the case of strain and pressure sensors, there is a trade-off between elastic properties and sensitivity. It is essential to find an optimum condition between these competing performance characteristics [29, 30]. It is also crucial to minimise the interference among various stimuli and address the effects of signal drift for accurate signal detection. Developing or improving detection mechanisms and algorithms, utilising machine learning techniques, or optimising circuit designs can help address these issues [16]. Significant research has investigated the use of graphene in developing wearable temperature sensors; however, several issues remain unresolved in current designs. Wearable temperature sensors encounter challenges related to stretchability, response time, and stability within the normal body temperature [19]. The annealing process of precursor materials, which consists of graphene platelets and nanowalls for producing rGrO, requires precise low-temperature conditions during manufacturing [31, 32]. Otherwise, it can decrease the mechanical strength of the synthesised rGrO. As a result, the overall performance of sensors can be adversely affected. Suspended graphene presents a viable alternative to nanocomposite-based sensors for the development of flexible thermistors. This is justified not only by its superior properties, including high elasticity, optical transparency, exceptional electron mobility, and excellent thermal and electrical conductivity [33], but also by the significant limitations of nanocomposite-based sensors, such as reduced electrical conductivity, structural degradation, poor long-term stability, and inconsistent performance [34, 35]. Consequently, this presents a promising approach to addressing these challenges and enhancing the functionality of flexible thermistors.

Further research is necessary to enhance the sensitivity, robustness, and longevity of graphene-based temperature sensors. One potential avenue for exploration is modifying the graphene band gap to customise the sensitivity of the developed prototypes. Incorporating semiconducting nanoparticles can also enhance accuracy and precision while reducing power consumption [19]. Choosing nanomaterials with low Young's modulus would improve the mechanical flexibility of the sensors. To enhance the washability of the resulting sensors, graphene could be combined with hydrophobic materials. Further studies are needed to strengthen the robustness of graphene-based sensors, particularly against repeated washings. Developing improved mechanisms or materials for more effective and flexible temperature sensors is therefore necessary.

5.3 Future Work

5.3.1 Multifunctionality

Graphene-based biomedical sensors should be multifunctional in their applications, rather than relying on a single sensor for a specific purpose. Numerous applications of graphene-based flexible electronics in monitoring human health and activity have emerged. These include the ability to track pulse, breathing, speech, body temperature, glucose, and lactate levels in sweat or tears, as well as ECG and EMG readings [14]. However, multifunctional wearable devices remain comparatively rare and deserve increased attention.

In 2021, Shirhatti et al. introduced a flexible wearable sensor featuring an extremely thin layer of graphene nanosheets on laser-engraved, circular, interdigitated gold electrodes [36]. They developed a custom laser-etching process to fabricate the sensor's circular, interdigitated electrodes. The functionalities of this multifunctional sensor were step counting, heart rate detection, motion monitoring, breath rate monitoring, body temperature sensing and dehydration detection.

Wang et al. proposed a multifunctional E-tattoo with healing ability [37]. This E-tattoo was developed by combining graphene, silk fibroin, and calcium ions. It was reported to monitor ECG, breathing, and temperature. Though some research has been conducted on the multifunctionality of wearable sensors, more attention is required in this sector because the advantage of using multifunctional sensors is that they can offer comprehensive monitoring and analysis capabilities while saving space and potentially reducing costs associated with deploying multiple single-function sensors. Their versatility and efficiency are essential in advancing healthcare technologies and enabling more holistic and precise patient care.

Graphene-based sensors are typically designed for single-use applications. Therefore, the oxidation issue of graphene is often overlooked. Oxidation becomes critical when it comes to devices intended for long-term usage. The presence of water molecules stabilises the O^{2-} anion and accelerates the oxidation process of graphene. Therefore, it is essential to develop and utilise an effective hydrophobic substrate to encapsulate the sensors from oxidation, thereby improving the durability and operational efficiency of the device [15].

5.3.2 Data Management and Artificial Intelligence

Creating a data management system integrated with machine learning for individuals that allows data collection and management from various graphene-based sensors or wearables across different body parts is another challenge. Furthermore, as medical information is confidential, developing a secure and private data processing system is important [12]. Developing a data management system integrated with artificial intelligence is crucial for collecting and analysing data generated by various sensors from different body areas. Additionally, it is essential to ensure a secure processing system to transfer data from sensors, thereby maintaining the confidentiality of medical information. Integrating the components necessary for creating graphene sensors will pose a challenge [12]. Developing biocompatible miniaturised parts will be essential to achieving fully wearable systems. Integrating big data analysis and machine learning techniques could enable systematic and intelligent processing, potentially transforming the traditional medical system and benefiting society [23]. There is a growing demand for integrated wearable sensing systems capable of monitoring multiple health-related indicators for future practical applications. Establishing such innovative systems requires the seamless integration of wearable sensors with essential components, including signal transmission, data collection and processing, wireless communication, power supplies, and feedback mechanisms, all while minimising size and optimising performance [14]. Additionally, intelligent systems designed to interact with the human body present considerable effects [14]. When combined with machine learning and big data approaches, graphene-based wearable sensors have the potential to meet medical and clinical needs. These sensors and their wearable applications will notably improve the quality of life, enabling diagnosis and treatment to occur within our daily routines [15].

5.3.3 *Other*

Although graphene is expected to exhibit long-term durability, the influence of mechanical strain on material degradation and the performance of graphene-based wearable sensors remains under-explored. Frequently, the consequences of integrating diverse biological samples are overlooked [12]. The incorporation of novel functional groups may augment intrinsic properties or introduce additional functionalities that enhance sensing capabilities [23]. Such factors must be carefully evaluated to improve the detection sensitivity of graphene biosensors. Moreover, in the development of integrated sensors for routine applications, it is essential to mitigate biofouling, including microbial interactions and other contaminants that may interfere with sensor outputs, leading to erroneous readings [12]. The long-term stability and reliability of fabricated sensors should be evaluated through real-world testing. Where necessary, robust encapsulation strategies must be developed to maintain sensor performance and reliability.

Graphene-based enzymatic biofuel cells (EBFCs) with self-powering capabilities show considerable promise for real-time biosensing applications, particularly in harnessing energy from biofluids like glucose or lactate [23]. Future investigations could prioritise expanding their utility to analyse diverse physiological fluids, including urine, semen, and mucus, to broaden diagnostic applicability, though current research primarily focuses on glucose-rich biofluids. Heteroatom doping, such as nitrogen or boron integration into graphene structures, enhances electron transfer efficiency and catalytic activity, yet scalable fabrication methods remain a critical challenge for widespread implementation [12]. Over time, such advancements in continuous, biocompatible monitoring systems may transform personalised healthcare by enabling early disease detection and improved health management. Rigorous efforts are needed to establish robust correlations between detected biomarkers and specific pathologies supported by standardised diagnostic frameworks to maximise their clinical utility. Concurrently, advancing protocols for data interpretation and therapeutic interventions will be essential to fully realise the potential of graphene-enabled wearable technologies [23].

5.4 Summary

Wearable biomedical sensors based on graphene show great promise due to their superior mechanical, chemical, thermal, and electrical properties. Graphene-based biosensors encounter persistent challenges, including ensuring biocompatibility, scaling up production, managing costs, addressing performance inconsistencies, and maintaining stability. Future research priorities should focus on developing multifunctional, self-powered systems with enhanced sensitivity and stability, while advancing cost-effective manufacturing techniques. The challenges and future research scope of graphene-based biomedical sensors presented in this chapter may guide research towards commercialisation and the graphene revolution in biomedical sensors.

References

1. H. Huang, S. Su, N. Wu, H. Wan, S. Wan, H. Bi, L. Sun, Graphene-based sensors for human health monitoring. Front. Chem. **7**, 399 (2019)
2. Q. Zheng, J.-H. Lee, X. Shen, X. Chen, J.-K. Kim, Graphene-based wearable piezoresistive physical sensors. Mater. Today **36**, 158–179 (2020)
3. H. Zhang, R. He, H. Liu, Y. Niu, Z. Li, F. Han, J. Li, X. Zhang, F. Xu, A fully inte- grated wearable electronic device with breathable and washable properties for long-term health monitoring. Sens. Actuators, A **322**, 112611 (2021)
4. S. Syama, P. Mohanan, Comprehensive application of graphene: emphasis on biomedical concerns. Nano-micro Lett. **11**, 1–31 (2019)
5. C. Liao, Y. Li, S.C. Tjong, Graphene nanomaterials: synthesis, biocompatibility, and cytotoxicity. Int. J. Mol. Sci. **19**(11), 3564 (2018)
6. E. Sawosz, S. Jaworski, M. Kutwin, A. Hotowy, M. Wierzbicki, M. Grodzik, N. Kurantowicz, B. Strojny, L. Lipin´ska, A., Chwalibog, Toxicity of pristine graphene in experiments in a chicken embryo model. Int. J. Nanomed. 3913–3922 (2014)
7. O. Akhavan, E. Ghaderi, A. Akhavan, Size-dependent genotoxicity of graphene nanoplatelets in human stem cells. Biomaterials **33**(32), 8017–8025 (2012)
8. B. Li, J. Yang, Q. Huang, Y. Zhang, C. Peng, Y. Zhang, Y. He, J. Shi, W. Li, J. Hu et al., Biodistribution and pulmonary toxicity of intratracheally instilled graphene oxide in mice. NPG Asia Mater. **5**(4), 44–44 (2013)
9. A. Dasgupta, J. Sarkar, M. Ghosh, A. Bhattacharya, A. Mukherjee, D. Chattopadhyay, K. Acharya, Green conversion of graphene oxide to graphene nanosheets and its biosafety study. PLoS ONE **12**(2), 0171607 (2017)
10. C.Z. Liao, H.M. Wong, H.M., K.W.K. Yeung, S.C. Tjong, The development, fabrication, and material characterization of polypropylene composites reinforced with carbon nanofiber and hydroxyapatite nanorod hybrid fillers. Int. J. Nanomed. 1299–1310 (2014)
11. K.W. Chan, C.Z. Liao, H.M. Wong, K.W.K. Yeung, S.C. Tjong, Preparation of polyetheretherketone composites with nanohydroxyapatite rods and carbon nanofibers hav- ing high strength, good biocompatibility and excellent thermal stability. RSC Adv. **6**(23), 19417–19429 (2016)
12. S.U. Singh, S. Chatterjee, S.A. Lone, H.-H. Ho, K. Kaswan, K. Peringeth, A. Khan, Y.-W. Chiang, S. Lee, Z.-H. Lin, Advanced wearable biosensors for the detection of body fluids and exhaled breath by graphene. Microchim. Acta **189**(6), 236 (2022)
13. J. Xie, Q. Chen, H. Shen, G. Li, Wearable graphene devices for sensing. J. Electrochem. Soc. **167**(3), 037541 (2020)
14. C. Wang, K. Xia, H. Wang, X. Liang, Z. Yin, Y. Zhang, Advanced carbon for flexible and wearable electronics. Adv. Mater. **31**(9), 1801072 (2019)
15. Y. Qiao, X. Li, T. Hirtz, G. Deng, Y. Wei, M. Li, S. Ji, Q. Wu, J. Jian, F. Wu et al., Graphene-based wearable sensors. Nanoscale **11**(41), 18923–18945 (2019)
16. M. Kong, M. Yang, R. Li, Y.-Z. Long, J. Zhang, X. Huang, X. Cui, Y. Zhang, Z. Said, C. Li, Graphene-based flexible wearable sensors: mechanisms, challenges, and future directions. Int. J. Adv. Manuf. Technol. **131**(5), 3205–3237 (2024)

17. H.C. Lee, W.-W. Liu, S.-P. Chai, A.R. Mohamed, A. Aziz, C.-S. Khe, N.M. Hidayah, U. Hashim, Review of the synthesis, transfer, characterization and growth mechanisms of single and multilayer graphene. RSC Adv. **7**(26), 15644–15693 (2017)
18. D. Boyd, W.-H. Lin, C.-C. Hsu, M. Teague, C.-C. Chen, Y.-Y. Lo, W.-Y. Chan, W.-B. Su, T.-C. Cheng, C.-S. Chang et al., Single-step deposition of high-mobility graphene at reduced temperatures. Nat. Commun. **6**(1), 6620 (2015)
19. A. Nag, R.B. Simorangkir, D.R. Gawade, S. Nuthalapati, J.L. Buckley, B. O'Flynn, M.E. Altinsoy, S.C. Mukhopadhyay, Graphene-based wearable temperature sensors: A review. Mater. Des. **221**, 110971 (2022)
20. K.-I. Jang, K. Li, H.U. Chung, S. Xu, H.N. Jung, Y. Yang, J.W. Kwak, H.H. Jung, J. Song, C. Yang et al., Self-assembled three dimensional network designs for soft electronics. Nat. Commun. **8**(1), 1–10 (2017)
21. Q. Guo, J. Koo, Z. Xie, R. Avila, X. Yu, X. Ning, H. Zhang, X. Liang, S.B. Kim, Y. Yan et al., A bioresorbable magnetically coupled system for low-frequency wireless power transfer. Adv. Func. Mater. **29**(46), 1905451 (2019)
22. M. Chung, G. Fortunato, N. Radacsi, Wearable flexible sweat sensors for healthcare monitoring: a review. J. R. Soc. Interface **16**(159), 20190217 (2019)
23. H. Zhang, R. He, Y. Niu, F. Han, J. Li, X. Zhang, F. Xu, Graphene-enabled wearable sensors for healthcare monitoring. Biosens. Bioelectron. **197**, 113777 (2022)
24. X. Wang, G. Sun, P. Routh, D.-H. Kim, W. Huang, P. Chen, Heteroatom-doped graphene materials: syntheses, properties and applications. Chem. Soc. Rev. **43**(20), 7067–7098 (2014)
25. C. Tan, X. Cao, X.-J. Wu, Q. He, J. Yang, X. Zhang, J. Chen, W. Zhao, S. Han, G.-H. Nam et al., Recent advances in ultrathin two-dimensional nanomaterials. Chem. Rev. **117**(9), 6225–6331 (2017)
26. M. Lafkioti, B. Krauss, T. Lohmann, U. Zschieschang, H. Klauk, K.v. Klitzing, J.H. Smet, Graphene on a hydrophobic substrate: doping reduction and hysteresis suppression under ambient conditions. Nano Lett. **10**(4), 1149–1153 (2010)
27. A. Pantelopoulos, N.G. Bourbakis, A survey on wearable sensor-based systems for health monitoring and prognosis. IEEE Trans. Syst. Man Cybern. Part C (Applications and Reviews) **40**(1), 1–12 (2009)
28. A.J. Bandodkar, I. Jeerapan, J. Wang, Wearable chemical sensors: present challenges and future prospects. Acs Sensors **1**(5), 464–482 (2016)
29. Y. Cheng, R. Wang, J. Sun, L. Gao, A stretchable and highly sensitive graphene-based fiber for sensing tensile strain, bending, and torsion. Adv. Mater. **27**(45), 7365–7371 (2015)
30. K.-Y. Chen, Y.-T. Xu, Y. Zhao, J.-K. Li, X.-P. Wang, L.-T. Qu, Recent progress in graphene-based wearable piezoresistive sensors: From 1d to 3d device geometries. Nano Mater. Sci. **5**(3), 247–264 (2023)
31. C. Zhou, X. Zhang, H. Zhang, X. Duan, Temperature sensing at the robot fingertip using reduced graphene oxide-based sensor on a flexible substrate. In: 2019 IEEE Sensors, pp. 1–4 (2019). IEEE
32. K. Gouda, S. Bhowmik, B. Das, Thermomechanical behavior of graphene nanoplatelets and bamboo micro filler incorporated epoxy hybrid composites. Mater. Res. Express **7**(1), 015328 (2020)
33. J. Lau, W. Bao Jr., J., Properties of suspended graphene membranes. Mater. Today **15**, 238–245 (2012). https://doi.org/10.1016/S1369-7021(12)70114-1
34. S. Godara, N. Sharma, Multifunctional applications of carbon nanotube–based polymer composites. In: Handbook of Carbon Nanotubes, pp. 1923–1936. Springer, Cham, Switzerland (2022)
35. E. Omanovi´c-Mikliˇcanin, A. Badnjevi´c, A. Kazlagi´c, M. Hajlovac, Nanocomposites: a brief review. Health Technol. **10**(1), 51–59 (2020)
36. V. Shirhatti, S. Nuthalapati, V. Kedambaimoole, S. Kumar, M.M. Nayak, K. Rajanna, Multifunctional graphene sensor ensemble as a smart biomonitoring fashion accessory. ACS Sensors **6**(12), 4325–4337 (2021)

37. Q. Wang, S. Ling, X. Liang, H. Wang, H. Lu, Y. Zhang, Self-healable multifunctional electronic tattoos based on silk and graphene. Adv. Func. Mater. **29**(16), 1808695 (2019)

If you have any concerns about our products,
you can contact us on
ProductSafety@springernature.com

In case Publisher is established outside the EU,
the EU authorized representative is:
Springer Nature Customer Service Center GmbH
Europaplatz 3, 69115 Heidelberg, Germany

Printed by Libri Plureos GmbH
in Hamburg, Germany